Managing Employees in Foodservice Operations

David K. Hayes
Jack D. Ninemeier

SENIOR EDITORIAL DIRECTOR	Justin Jeffryes
EXECUTIVE EDITOR	Todd Green
EDITORIAL ASSISTANT	Kelly Gomez
SENIOR MANAGING EDITOR	Judy Howarth
PRODUCTION EDITOR	Mahalakshmi Babu
COVER PHOTO CREDIT	© andresr/Getty Images

This book was set in 9.5/12.5 STIX Two Text by Straive™.
Published by John Wiley & Sons, Inc., Hoboken, New Jersey.
Published simultaneously in Canada.
This book is printed on acid-free paper.

Founded in 1807, John Wiley & Sons, Inc. has been a valued source of knowledge and understanding for more than 200 years, helping people around the world meet their needs and fulfill their aspirations. Our company is built on a foundation of principles that include responsibility to the communities we serve and where we live and work. In 2008, we launched a Corporate Citizenship Initiative, a global effort to address the environmental, social, economic, and ethical challenges we face in our business. Among the issues we are addressing are carbon impact, paper specifications and procurement, ethical conduct within our business and among our vendors, and community and charitable support. For more information, please visit our website: www.wiley.com/go/citizenship.

ISBN: 978-1-394-20841-8 (PBK)

Library of Congress Cataloging-in-Publication Data:

Names: Hayes, David K., author. | Ninemeier, Jack D., author. | John Wiley
    & Sons, publisher.
Title: Managing employees in foodservice operations / David K. Hayes, Jack
    D. Ninemeier.
Description: Hoboken, New Jersey : Wiley, [2024] | Includes index.
Identifiers: LCCN 2023043152 (print) | LCCN 2023043153 (ebook) | ISBN
    9781394208418 (paperback) | ISBN 9781394208432 (adobe pdf) | ISBN
    9781394208449 (epub)
Subjects: LCSH: Food service—Personnel management.
Classification: LCC TX911.3.P4 H395 2024 (print) | LCC TX911.3.P4 (ebook)
    | DDC 647.95068—dc23/eng/20231017
LC record available at https://lccn.loc.gov/2023043152
LC ebook record available at https://lccn.loc.gov/2023043153

The inside back cover will contain printing identification and country of origin if omitted from this page. In addition, if the ISBN on the back cover differs from the ISBN on this page, the one on the back cover is correct.

Contents

Preface

Professional foodservice operators in all segments of the industry recognize that providing high-quality menu items and excellent service is essential to their long-term success. These same operators must recognize that providing excellence in product and service quality is impossible without the efforts of a highly qualified, well-trained, and committed team of employees.

The purpose of this book is to teach foodservice operators what they must know, and do, to attract, train, and retain work teams that allow the operators to reach their financial goals, while at the same time allowing employees the ability to achieve their own personal and professional goals. A major premise of this book is that the best interests of foodservice employers are nearly always in alignment with the best interests of their employees.

Many segments of the foodservice industry have historically faced challenges in securing the needed number of qualified employees. The COVID-19 pandemic of the early 2020s, however, which affected the foodservice industry in many ways, lead to a significant decline in the workforce available to many foodservice operators. As a result, even those operators who had not faced serious labor shortage challenges in the past were forced to reassess the importance of employees to the successful operation of their businesses. As a result, the successful management of employees has now taken on more importance than ever before.

It is vital to understand that, in the United States, foodservice operators have a great deal of freedom in how they manage their employees and their businesses. That freedom is not, however, unlimited. In this book readers will learn that all foodservice operators must follow applicable federal, state, and local laws and ordinances. They will also learn about important legislation related to protecting workers' health as well as the many financial benefits of doing so.

There is no question that, in today's very tight and competitive labor market, obtaining and retaining an efficient workforce is one of a foodservice operator's most significant challenges. In this book, readers will discover that the identification and assembly of an effective foodservice work team begins with proper job design. How foodservice operators initially design their jobs can have a significant impact on the overall appeal of these jobs. This book directly addresses how jobs can be designed in ways that make them as attractive as possible to potential employees.

All foodservice operators recognize that their "labor costs" have a direct impact on their profitability; unfortunately, that can lead some operators to think that labor costs that are "low" are more desirable than those that are "high." This approach is not only wrong, it is overly simplistic. An important premise of this book is that food service operators want to *optimize* their labor expense, not minimize it. An optimal labor cost is one that allows foodservice operators to achieve their financial objectives, while at the same time ensuring employees feel their jobs are highly desirable and satisfying. The best foodservice operators recognize that competitive compensation packages significantly impact their ability to attract and retain highly qualified workers. This book directly addresses how employee compensation packages can be used to attract the number, and quality, of employees needed to operate a foodservice business successfully.

This book is distinctive in that it addresses those employee management issues that are unique to the foodservice business. Among the many key topics addressed in this book are:

✓ The importance of leadership in employee management
✓ How to ensure compliance with employment law
✓ The importance of maintaining a safe and healthy workplace
✓ How to establish fair employment policies and procedures
✓ How to recruit new team members in tight labor markets
✓ How to interview and select new team members
✓ How to create an employee orientation program that optimizes retention
✓ How to effectively train new and existing team members
✓ How to create an employee compensation program that attracts high-quality employees
✓ How to schedule staff for maximum effectiveness and employee retention
✓ How to create an employee appraisal and progressive discipline program
✓ How to assess an operation's labor usage effectiveness
✓ How to manage and control an operation's total labor-related costs

Most importantly, all of these topics and more are addressed in ways specific to foodservice operations of all sizes and in all industry segments. The ability to develop a productive workforce is essential to the success of every foodservice operation regardless of whether the operation is a food truck, coffee kiosk, ghost restaurant, quick service restaurant, full-service operation, or a non-commercial facility.

Readers of this book will quickly find that the content of this book is essential to the successful management of their own operations, and they will also find that the information in each chapter has been carefully selected to be easy-to-read, easy-to-understand, and easy-to-apply.

Book Features

In addition to the essential employee management information it contains, special features were carefully crafted to make this learning tool powerful but still easy to use. These features are:

1) **What You Will Learn** To begin each chapter, this very short conceptual bulleted list summarizes key issues readers will know and understand when they complete the chapter.
2) **Operator's Brief** This chapter opening overview states what information will be addressed in the chapter and why it is important. This element provides readers with a broad summation of all important issues addressed in the chapter.
3) **Chapter Outline** This two-level outline feature makes it quick and easy for readers to find needed information within the body of the chapter.
4) **Key Terms** Professionals in the foodservice industry often use very special terms with very specific meanings. This feature defines important (key) terms so that readers will understand and be able to speak a common language as they discuss issues with their colleagues in the foodservice industry. These key terms are also listed at the end of each chapter in the order in which they initially appeared.
5) **Find Out More** In a number of key areas readers may want to know more detailed information about a specific topic or issue. This useful book feature gives readers specific instructions on how to conduct an Internet search to access that information and why it will be of importance to them.
6) **Technology at Work** Advancements in technology play an increasingly important role in many aspects of foodservice operations. This feature was developed to direct readers to specific technology-related Internet sites that will allow them to see how advancements in technology can assist them in reaching their operational goals.
7) **What Would You Do?** These "mini" case studies located in every chapter of the book take the information presented in the chapter and uses it to create a true-to-life foodservice industry scenario. They then ask the reader to think about their own response to that scenario (i.e., *What Would You Do?*).

This element was developed to help heighten a reader's interest and to plainly demonstrate how the information presented in the book relates directly to the practical situations and challenges foodservice operators face in their daily activities.

8) **Operator's 10-Point Tactics for Success Checklist** Each chapter concludes with a checklist of tactics that can be undertaken by readers to improve their operations and/or personal knowledge. For example, in a chapter of the book addressing "Interviewing and Selecting New Team Members," one point in that chapter's 10-Point Tactics for Success Checklist is:

Operator understands the advantages and potential drawbacks associated with utilizing background checks and references in the new employee selection process.

Instructional Resources

This book has been developed to include learning resources for instructors and for students.

To Instructors

To help instructors (and corporate trainers!) effectively manage their time and enhance student-learning opportunities, the following resources are available on the instructor companion website at www.wiley.com/go/hayes/managingemployeesfoodservice.

✓ Instructor's Manual that includes author commentary for "What Would You Do?" mini case-study questions.

✓ PowerPoint slides for instructional use in emphasizing key concepts within each chapter.

✓ A 100-item Test Bank consisting of multiple-choice exam questions, their answers, and the location within the book from which the question was obtained. The test bank is available as a print document and as a Respondus computerized test bank. *Note*: **Respondus** is an easy-to-use software program for creating and managing exams that can be printed to paper or published directly to Blackboard, WebCT, Desire2Learn, eCollege, ANGEL, and other e-learning systems.

To Students

Learning about the management of employees in foodservice operations will be fun. That's a promise from the authors to you. It is an easy promise to make and keep because working in the foodservice industry is fun. And it is challenging. However, if you work hard and do your best, you will find that you can master all of the important information in this book.

When you do, you will have gained invaluable knowledge that will enhance your skills and help advance your own hospitality career. To help you learn the information, in this book, online access to over 225 PowerPoint slides is available to you. These easy-to-read tools are excellent study aids and can help you when taking notes in class.

Acknowledgments

Managing Employees in Foodservice Operations has been designed to be the most up-to-date, comprehensive, technically accurate, and reader-friendly learning tool available to those who want to know how to attract, train, and retain an effective team of dedicated foodservice employees.

The authors thank Catriona King of Wiley for initially working with us to develop the idea for a series of practical books that would help foodservice operators of all sizes more effectively manage their businesses. She was essential in helping conceptualize the need for this book as well as all of the other books in this five-book Foodservice Operations series. The five titles in the "Foodservice Operations" series are:

Managing Employees in Foodservice Operations
Marketing in Foodservice Operations
Accounting and Financial Management in Foodservice Operations
Cost Control in Foodservice Operations
Successful Management in Foodservice Operations

We would also like to thank the external reviewers who gave so freely of their time as they provided critical industry and academic input on this series. To our reviewers, Dr. Lea Dopson, Gene Monteagudo, Isabelle Elias, and Peggy Richards Hayes we are most grateful for your comments, guidance, and insight. Also, thanks to Biloxi Mississippi's Michael T. Kavanagh, who was a technological friend indeed, when we were most in need!

Books such as this require the efforts of many talented specialists in the publishing field. The authors were extremely fortunate to have Todd Green, Judy Howarth, and Kelly Gomez at Wiley as our publication team. Their efforts and creativity went far in helping the authors present the book's material in the best and clearest possible form.

Finally, the authors would like to thank the many students and industry professionals with whom we have interacted over the years. We sincerely hope this book allows us to give back to them as much as they have given to us.

David K. Hayes, Ph.D.
Jack D. Ninemeier, Ph.D.

Dedication

The authors are delighted to have the opportunity to dedicate this book, and the entire Foodservice Operations series, to two outstanding and unique individuals.

Brother Herman Zaccarelli

Brother Herman E. Zaccarelli, C.S.C., passed away in 2022 at the Holy Cross House in Notre Dame, Indiana. His professional work included many projects for the hospitality industry, and he published several books and hundreds of articles for numerous trade publications over many years. Among numerous accomplishments, Herman founded Purdue University's Restaurant, Hotel, and Institutional Management Institute in 1976. Later, he served as Director of Business and Entrepreneurial Management at St. Mary's University in Winona, Minnesota.

A lifelong learner, at the age of 68, Brother Herman retired to Florida where he earned a Bachelor's degree in Educational Administration and a Master's degree in Institutional Management.

Herman's ideas and concepts have been widely adopted in the hospitality industry, and he assisted many young educators including the authors of this book series. He will be remembered as a colleague with creative ideas who provided significant assistance to those studying and managing in the hospitality industry. Herman was especially helpful in discovering and addressing learning opportunities for Spanish speaking students, educators, and managers throughout the United States and around the world.

Dr. Lea R. Dopson

A lifelong friend, advisor, and colleague, as well as an outstanding author herself; at the time of her untimely passing, Lea served as President of the International Council on Hotel, Restaurant, and Institutional Education (ICHRIE) and Dean of the prestigious Collins College of Hospitality Management (Cal Poly Pomona).

Lea was a dedicated hospitality professional and a fierce advocate for hospitality students at all levels. Those who knew her were continually in awe of her intelligence and humility.

It was especially fitting that Lea was named as a recipient of the H.B. Meek Award. That award is named after the individual who started the very first hospitality program in the United States (at Cornell University). Selected by the recipient's peers, it goes not to the most outstanding academic professional working in the United States but to the most outstanding academic professional in the entire world. That was Lea.

While she is dearly missed, her inspiration goes on everlastingly in the works of the authors.

1

Leading the Foodservice Team

What You Will Learn

1) Why Foodservice Operators Must Be Effective Leaders
2) How Operators Best Serve Guests and Employees
3) Important Leadership Practices
4) Special Concerns of Foodservice Leaders

Operator's Brief

Foodservice operators invest significant money and effort as they develop and maintain their organizations. However, they may sometimes face significant difficulties in recruiting and retaining qualified foodservice employees. Sometimes operators are unaware that one reason this situation occurs is that the work atmosphere does not help the staff to attain their own financial and other career goals.

In this chapter, you will learn that the best foodservice operators ensure that the work culture addresses their employees' concerns as well as those of the guests.

As a foodservice operator, you will be the leader of your organization and there is no one leadership style that works best for every operator. Instead, experienced foodservice operators consider leadership style alternatives as they interact with their staff members. They also develop mission statements and implement them by demonstrating their core values whenever possible.

Guests are concerned about the quality of products and services they buy. Employees are concerned about fair compensation for the work they do and about receiving personal respect from their leaders. The list of staff concerns can be long and addressing them requires a strong foundation of information incorporated into planning, organizing, staffing, directing, controlling, and evaluating employees' work activities.

In this chapter, you will learn why foodservice operators must be effective leaders, how you can successfully address both your guests' and employees' concerns, and important leadership practices to employ as you do so.

CHAPTER OUTLINE

Foodservice Operators as Leaders
Serving Guests and Employees
    Leadership Styles
    Mission Statements and Core Values
Important Leadership Practices
    Act the Way Employees Should Act
    Share Mission Statement and Core Values with Staff Members
    Challenge Existing Processes and Procedures
    Enable Employees with Trust and Delegation
    Show Appreciation for Individual Excellence
Special Concerns of Foodservice Leaders
    Employee Empowerment
    Employee Team Development
    Benefits of Effective Teamwork
    Characteristics of Effective Team Leaders
    Cross-functional and Self-directed Work Teams
    Ethics

Foodservice Operators as Leaders

The best foodservice operators know they must effectively lead their work teams as they manage their businesses. What is **management**? A simple definition suggests that "management is the process of using available resources to attain desired goals."

A special concern about an organization's resources is that they are almost always limited, and few, if any, operators have all the resources that are desired as operating goals are prioritized and pursued.

Key Term

Management: The process of planning, organizing, directing, controlling, and evaluating the financial, physical, and human resources of an organization to reach its goals.

Examples of resources in a foodservice operation include money, facilities, equipment, employees, available time, and the work procedures that are in place. Other resources may include energy, inventories, and specialized items such as recipes and even specialized cooking techniques. Since resources are limited, however, foodservice operators must make good decisions about how they will be used.

There are numerous ways to view the principles of good management and the management process. One way is to consider the five primary functions that are essential to effectively manage an organization. These are shown in Figure 1.1.

Planning: This initial step in the management process addresses the creation of goals and objectives. This is the first step in the management process because it

Figure 1.1   The Management Process

identifies precisely what an organization wants to achieve. It is important to use current and pertinent information and, when possible, the participative input from staff members who are affected by the plans. Plans must be flexible, and they must be implemented, evaluated, and changed when appropriate.

Organizing: After its objectives have been identified, an organization must ensure it has the funding, staff, equipment, and raw materials needed to achieve its objectives. These business **assets** must then be arranged (organized) in a way that optimizes the organization's ability to achieve its objectives.

Key Term

Asset (business): Property that is used in the operation of a business including money, real estate, buildings, inventories, and equipment.

Directing and Leading: This important management function addresses the task of telling and showing all staff members exactly what is expected of them. When given clear directions, all staff members will know the important roles they will play in helping the organization achieve its objectives.

The best foodservice operators strive to be good leaders. While this is an easy goal to state, it is often much more difficult to achieve. What makes a good leader? While entire books (and bookshelves!) address this key question, a short list of effective leadership traits for foodservice operators can be developed, and they include:

✓ Having a good understanding of the operation's values and being able to translate these values into effective practices.
✓ Having an objective and measurable "picture" of the desired future of the operation.

✓ Helping others to develop the knowledge and skills needed to attain the operation's vision. This is done, in part, through effective employee selection, orientation, training and follow-up evaluation, and coaching activities.

✓ Developing a team of staff members who are committed to the operation's success.

✓ Achieving a reputation for quality that enables the foodservice operation to increasingly meet, and exceed, their guests' expectations.

Controlling: By continually assessing the work processes and procedures they have implemented, foodservice operators can better identify situations that might prevent their organizations from meeting objectives. In the foodservice industry, important processes that must be controlled include those activities related to purchasing, receiving, storing, preparing, and serving menu items.

Control is an important responsibility of every foodservice operator. This task enables an operator to compare planned financial results with actual operating results. This analysis helps to indicate where, if at all, corrective actions might reduce variances between planned and actual financial results.

There are several steps involved in the control process, and each step can help ensure that the financial goals of an operation will, in fact, be attained. Operators forecast their sales levels, and then they develop cost control systems. This several step process can be summarized as:

1) Forecasting expected revenue and costs
2) Measuring actual operating results
3) Comparing expected and actual operating costs
4) Taking corrective cost-related actions as needed
5) Evaluating the effectiveness of the corrective actions

Find Out More

Many of the goals foodservice operators have for themselves and their businesses involve the effective control of operating costs and optimizing operating profits.

The effective control of costs in a foodservice operation requires extensive knowledge and skill in a variety of areas. While the final chapter of this book addresses the control of costs related to the staff that are needed to serve an operation's guests, there are other costs that must be controlled including the costs of food and beverage products, preparation costs, and guest service costs.

While there are many publications addressing "Cost Control" in foodservice operations, one of the best and most up-to-date cost control resources available to foodservice operators is *Cost Control in Foodservice Operations* written by Drs. David Hayes and Jack Ninemeier and published by John Wiley.

To learn more about the content and availability of this extremely valuable publication, go to www.wiley.com, enter "*Cost Control in Foodservice Operations*" and review the results.

Evaluating: This final management activity requires an organization to assess its current performance and compare that performance to that which was planned. If significant differences exist, the organization must determine the reason for the differences and then either change its objectives or revise the methods used to achieve them.

Although this book addresses each of the five key management processes, its primary focus is on directing and leading. Specifically, it addresses the directing and leadership skills foodservice operators must know and apply as they obtain, train, retain, and motivate outstanding work teams.

Serving Guests and Employees

Foodservice operators must be knowledgeable about many things. However, some critically important issues can be addressed in similar ways because they benefit both guests and employees.

Figure 1.2 shows a planned activities model that illustrates foodservice operators must focus on both guests and employees to achieve their organizational goals. Both guests and employees are important **stakeholders** of a foodservice operation, and it is critical that operators know and can address the major concerns of these two groups.

Key Term

Stakeholders: Groups, individuals, and organizations that have a vested interest in the decision-making and activities of a business or organization.

What do guests and employees expect from their interactions with a foodservice operation? One basic issue of concern to guests is their desire for **value**. In addition, guests want to know that their food safety concerns are addressed, that they will not be injured or become ill during their dining experience, and, increasingly, that the menu items they purchase are healthy and nutritious.

Key Term

Value: The amount paid for a product or service compared to the buyer's view of what they received in return.

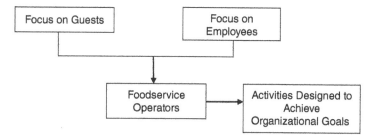

Figure 1.2   Foodservice Operator's Activity Planning

What concerns are important to employees? While they vary between specific staff members, examples of employee-centered needs typically include fair compensation including benefits for the work that they do, personal respect from their supervisor, and potential job advancement opportunities. Other employee-specific interests may include a flexible work schedule, a positive team atmosphere (employees helping each other!), job security, and, increasingly, diversity/inclusiveness.

The best foodservice operators know that while they must properly serve their guests, they must properly serve their work team members as well. It is easy to understand that, if guests are not properly served, the operation will not remain in business and employees will lose the opportunity to work!

Alternatively, if enough highly trained and professional staff are not available to properly service guests, an operation may not be able to achieve the revenue levels needed to stay in business. Increasingly, in today's labor market, foodservice operators are finding that obtaining and retaining qualified staff is just as much of a challenge as is attracting guests.

While there is a great deal of material available about how foodservice operators can best attract guests, the focus of this book is on how foodservice operators can best recruit and retain a highly qualified workforce.

Leadership Styles

The best foodservice operators lead their work teams. A leader is one who "gets things done by working with people." Anyone who manages or supervises people must possess leadership skills to be successful. In the not-too-distant past, leaders could simply tell staff members what to do. Then, when the assignment was successfully completed, the supervisor was "successful" and was considered to be a good leader.

Today, this strategy is much less effective, and a leader's ability to dictate has been replaced with the need to influence employees and to facilitate required work. The role of a foodservice operations leader has changed from one of being a "task master" to one of facilitating by providing employees with resources, guidance, and direction.

It is important to understand that there is not one "best" type of leadership style. In fact, those who study leadership have identified several different leadership styles, and their findings suggest that the best leaders can modify their leadership style to the appropriate situation. While there are several ways to define leadership styles, one good way is to classify basic leadership styles as one of the following:

Autocratic
Bureaucratic
Democratic
Laissez-faire

Autocratic

Autocratic leaders like to make decisions and solve problems without getting information from affected work team members. They give instructions unilaterally and expect employees to follow them. Sometimes these supervisors use a structured set of awards to encourage employees to follow orders and then discipline employees when orders are not followed.

Foodservice team members working under these supervisors can easily become dependent on their autocratic leaders. They may act only under the leader's supervision and direct instruction and have little opportunity to make decisions in the normal course of their work.

Autocratic leaders may demonstrate a genuine concern for guests that is exhibited in their concern for food and beverage production and service methods, but most often demonstrate a lessened concern about the needs and desires of their work team members.

Bureaucratic

Bureaucratic leaders "manage by the book" with an emphasis on enforcing the rules. Problems not addressed by the rules are typically referred to higher levels of management or ownership.

The bureaucratic leader is many times more of a police officer than a leader because the emphasis is on rules—not cooperation. As a result, following the rules often becomes more important than the team members' feelings and work results.

Democratic

Democratic leaders want to include team members in the decision-making activities that affect them. These leaders focus on subordinates and the role they play in the foodservice operation, and democratic leaders search for ways to help work team members find satisfaction in their jobs.

Democratic leaders are typically well-liked by their team members because they exhibit true concern for the team members. There are, however, some potential disadvantages to the democratic leadership approach. For example, decision-making is likely to take longer and mistakes are more likely to occur. Also, once this style has been used in a foodservice operation it may be difficult for team members to accept another leadership approach (e.g., when a new team leader displays a different leadership style).

Laissez-faire

Laissez-faire is a French term that translates to "leave alone." Laissez-faire leaders use a "hands-off" approach and do as little supervising and coaching as possible. Their foodservice team members make decisions with little input from their leaders. This approach can be successful but typically only with highly motivated and experienced foodservice staff members.

Figure 1.3 provides a summary of the leadership styles of various foodservice operators and the targeted team members for each style.

Type of Leadership	Leadership Overview	Target Team Members for This Leadership Style
Autocratic	This leader retains as much power and decision-making authority as possible. The leader makes decisions without consulting work team members. Orders are given to, and must be obeyed by, workers without allowing for their input.	New team members who must quickly learn work tasks; difficult to supervise employees who do not respond to other styles; and temporary employees.
Bureaucratic	This leader supervises "by the book." The emphasis is on doing things specified by rules, policies, regulations, and standard operating procedures (SOPs). This leader relies on higher levels of ownership or management to resolve problems not addressed by the rules.	Team members who must follow set procedures and those working with dangerous equipment or under other special circumstances.
Democratic	This leader involves work team members in aspects of the job that directly affect them. Team members' input is solicited and encouraged. Team members participate in the decision-making process and are delegated the authority to make some decisions.	Team members with high levels of skill or extensive experience; employees who need to agree to significant changes in work assignments; team members who want to voice complaints; and employee groups with common problems.
Laissez-faire	This leader maintains a "hands-off" leadership approach and delegates much decision-making authority. They give little direction and allow extensive levels of work team member freedom.	Highly motivated work team members and specialists with high levels of experience and/or technical skills.

Figure 1.3 Leadership Styles of Foodservice Operators

Serving Guests and Employees | 9

Technology at Work

Every foodservice operator is a unique individual, and many foodservice operators find that their own leadership style is a blend of styles with a specific style being used in specific situations.

Most leaders have a dominant leadership style that they are most comfortable using. As a result, it can be a good idea for you to identify your basic, or dominant, leadership style.

Fortunately, there are numerous free-to-use leadership style self-quizzes on the Internet that can help identify the style you will naturally lean toward, and that you might find helpful in understanding your own leadership characteristics.

To take one or more of the free leadership style tests available on the Internet enter, "free leadership style quiz" in your favorite search engine and take a quiz.

After you have finished taking the quiz, consider the following:

1) Were you surprised by the results?
2) Do you feel the quiz accurately assessed your leadership style?
3) Would you like to make any changes to your basic leadership style?

Ideally, a foodservice operator will likely know the best way to direct and lead the efforts of each staff member. However, this becomes very difficult when, for example, there are high-**employee turnover rates**, and many employees are replaced. As this occurs, there is likely to be an insufficient number of previous interactions to understand the best approaches to manage specific employees.

Experienced operators know they can best direct employees in different situations if they have some knowledge about how an employee will react to different leadership styles. Without this knowledge, an operator may only utilize the leadership approach with which they are most

Key Term

Employee turnover rate: The proportion of total employees in an operation replaced during a specific time period.

The formula used to calculate employee turnover rate is:

$$\frac{\text{Number of employees separated}}{\text{Number of employees in workforce}} = \text{Employee turnover rate}$$

familiar, and this may or may not yield successful interactions with the staff member.

It is unlikely that any operator consistently uses one of the above identified leadership styles in every situation and with every staff member. The decision of which approach is appropriate is ideally made based on a leader's ability to recognize and use the "best" leadership approach based on the situation.

For example, a new employee trained for an inadequate time before being left alone without any assistance would likely become frustrated with a "hands off" leadership approach. A very experienced staff member, however, who is consistently and harshly "reminded" by their supervisor about how to perform simple job tasks would likely become frustrated as well. In both situations, the best leadership approach involves providing a foundation of job-related information, respecting employees for the questions they ask, and thinking about the importance of both guests and employees as leadership decisions are made.

Mission Statements and Core Values

Regardless of the leadership style utilized, a foodservice operator must first understand their organization's **mission statement** and its core values to effectively lead a work team. A mission statement answers questions such as "Why do we exist?", "How will we be successful?", and "How will we be able to add proper value to our guests' interactions with us?"

Key Term

Mission statement: A concise explanation of the organization's reason for existence that describes the organization's purpose and its overall intention.

Practical and valuable mission statements are critically important, and they do not need to be complex. For example, a foodservice operation's mission statement might be stated as:

> *We are dedicated to providing exceptional service for our guests and a professional and respectful work environment for our staff members. Our guests will benefit from the excellence of team members who consistently address the challenges and demands of our dynamic and remarkable foodservice operation.*

Note that this sample mission statement mentions employees in two different places. This is reasonable because qualified employees are necessary to deliver the products and services that will yield an exemplary experience for the guests. Unfortunately, some foodservice operations do not have mission statements. Further, many of the properties that do have these tools do not mention employees even though the employees are those who will best ensure that the mission statement goals are addressed and attained in the interactions with the operation's guests.

Technology at Work

Regardless of the specific language used, a mission statement should always be an action-based statement of an organization's purpose and how it will serve its guests and employees. This most often includes a short description of the company (what it does) and its objectives. Therefore, a mission statement is actually a short summary of the foodservice operation's purpose.

Despite their importance, some foodservice operators have difficulty developing a useful mission statement because they simply do not know where to start. This problem is common enough that many entities now provide templates for mission statements that can be adapted for use by foodservice operators who want to create this important tool.

To view information about mission statement development and to see some samples that can be adapted for your own use, enter "sample mission statements for restaurants" in your favorite search engine and view the results.

Core values represent the fundamental beliefs held by a foodservice operation. These planning tools help implement mission statements because they suggest how all owners, operators, and employees should act and what behaviors are important. They can also specify a foodservice operation's morals, priorities, and commitments.

Core values should be adopted by every member of management and every employee at every organizational level within a foodservice operation. As a result, employees will know their roles in the organization's present and future, and every employee should understand that "It is not I or me; rather it is us!" Sample core values directly impacting employees might address topics such as:

Key Term

Core Values: Deeply ingrained principles that guide an organization's actions. They are fundamental beliefs about a foodservice operation that dictate behavior and help its members understand what is considered right or wrong.

✓ All decisions will be made in the context of the highest level of professional ethics.
✓ We will genuinely respect our employees through all our words and actions.
✓ Our emphasis on quality will be never-ending.
✓ All employees are eligible for additional training that is available.
✓ Our operation promotes from within.
✓ Employees will be empowered to make decisions that yield products and services that consistently provide excellent value to guests.

What Would You Do? 1.1

"I'm not saying you should say anything, I'm just saying they're making me very uncomfortable," said Gina, to her supervisor Ray.

Gina was a young female server at Spagnola's Italian restaurant. The Spagnola's operation consisted of a 125-seat dining room and a 45-seat cocktail lounge. Gina was the server working tonight in the lounge, and she was talking to Ray about a table of six young men who were in the lounge waiting for their table in the dining room.

"How are they making you uncomfortable?" asked Ray.

"It's their comments," replied Gina, "They keep commenting on my uniform and telling me how good I look in it. One of them also asked me what time I was getting off tonight. They haven't actually touched me or anything, but their comments have gotten more personal and suggestive over the past 15 minutes or so that I have been waiting on them."

Assume you were Ray. What would you think about Gina's suggestion that you should "not say" anything to this group? What message would you be sending Gina and the rest of your service staff if you elected not to talk to the group? What message would you likely be sending to your service team if you did elect to talk to this group?

Important Leadership Practices

Foodservice operators should consistently practice several important leadership principles as they direct and lead staff members. These principles relate to sharing and strengthening an operation's core values, and they represent an integral way that an operator should interact with staff members. Some important leadership practices for leaders in all foodservice operations are addressed in the following sections.

Act the Way Employees Should Act

The actions of a leader are far more important than words, and these actions are judged by employees when they consider how serious their leaders are about what they say. Obviously, a leader's words and actions must be consistent.

There are no unimportant tasks in an organization's efforts at attaining consistency between words and actions. To consistently meet this standard, leaders must be sincere about what they say. They must also pay attention to details and remember that simple actions can make a difference.

Consider, for example, a foodservice operator who is working side-by-side with an employee discussing how core values must dictate decision-making. What would that employee think if they walked past the operator's office with an open door and hears the operator instructing the operation's buyer to start purchase more products from a specific vendor because the vendor "will give me a nicer holiday gift than competitors!"

It is important to recognize that foodservice operators earn their right to lead by direct involvement and action. It is true that "people first follow the person and then the plan." A leader's personal and business behavior wins respect from staff members, and leaders gain commitment and achieve higher standards when they model the behavior expected from others.

Share Mission Statement and Core Values with Staff Members

The best foodservice leaders imagine exciting and attractive futures for their organizations. They have an absolute and total belief in and are confident that staff members also have an interest in and the abilities to make extraordinary things occur. They know that one cannot "command" commitment but, rather, operators can only inspire it.

Operators should also clearly understand that their employees also have personal dreams, needs, values, hopes, aspirations, and interests, and operators should attempt to merge these interests with those of the foodservice operation.

For example, a talented employee may want to own their own foodservice operation someday. An operator who knows this can, on an ongoing basis, discuss current challenges addressed by the operation and offer suggestions to the employee. Would sharing this information create future competition? Maybe, but it is also possible that the operator's mentoring will preface a promotion for the employee who might someday become the manager of the operator's second property!

Challenge Existing Processes and Procedures

Experienced foodservice operators recognize there are almost always better ways to do things. Therefore, they continually seek and accept the challenge of "changing for the better," and they are alert to ways that will help to innovate and improve. They know that positive change comes more from "listening" than from "telling."

It is important to challenge the system to adopt new products, processes, services, and systems. Operators should recognize that change normally arises through small steps and successes, and problem solutions can help to improve operations. Successful operators foster collaboration with staff members by seeking their ideas and by promoting cooperative goals and building trust. They understand their staff members interact with other stakeholders including fellow employees, guests, and vendors, and all parties can often help to identify challenges and offer solutions to address them.

Enable Employees with Trust and Delegation

Effective leaders foster collaboration by promoting cooperative goals and by building trust between themselves and their individual team members. As they do this, they share power and decision-making and successful leaders say "we" more often than "I." They also recognize that leadership is a team effort, and they enable others to act because they foster collaboration, build trust, and engage all staff members as they make process changes and revise applicable operating procedures.

The best leaders **delegate** tasks to staff members who are prepared to assume them. These successful employees are made to feel strong and capable, and they become committed to the foodservice operation. Those who are trusted are given more decision-making opportunities, power, and information, and they are those most likely to use their creative energies to produce positive results.

Key Term

Delegate: To assign authority to subordinates to do work and make decisions normally made by someone higher at a higher organizational level.

Show Appreciation for Individual Excellence

Personal credibility is the foundation of leadership, and operators should recall two relevant old sayings: "If you can't believe the messenger, you won't believe the message" and "Leadership is a relationship between those who want to lead and those who choose to follow."

Encouragement, including recognition, can motivate many employees. This might come from a dramatic gesture (for example, a special article posted on an operation's website discussing a new menu item suggested by an employee). Alternatively, only a simple action (for example, thanking an employee at a meeting for a suggestion to revise the inventory process) might be involved, but both of these actions can motivate team members.

Many management observers have noted that the primary success factor in leadership is the operator's relationship with staff members. Leadership can be a "back-and-forth" process between those who aspire to lead and those who elect to follow. When those who aspire to lead address their employee's expectations, staff members are more likely to recognize their leaders to be credible, and then staff are more likely to show and tell others they are part of the team.

Motivated employees generally have a strong sense of team spirit. A leader's credibility also improves when employees believe their own personal values are consistent with those of their employer. These employees feel attached and committed to the organization because they have a strong sense of ownership with it. When leadership is a relationship founded on trust and confidence, employees are more likely to take controlled risks, make changes, and help to ensure a foodservice operation remains able to take the steps necessary to properly service its guests.

In contrast, when employees perceive their leaders to have low credibility, staff are more likely to produce only to the extent they are forced to. They are also more likely to be motivated primarily by money. Also, if staff members speak positively about their employer in public, they may still criticize the foodservice operator privately. These staff are among the employees who will continually look for another employer because they feel under-supported and under-appreciated in their present jobs.

Special Concerns of Foodservice Leaders

Foodservice operators continue their employee-related concerns as they plan empowerment programs and form employee groups into organized teams and manage them, and they also address important ethical implications common to their positions.

Employee Empowerment

Employee **empowerment** involves authorizing employees to make discretionary decisions within their areas of responsibility. Empowerment results from a leader's efforts to fully involve team members in the decision-making process if they (the team members) wish to assist with this process.

Many experienced operators will remember that it was once customary for management to (a) make all decisions regarding almost every

Key Term

Empowerment (employee): An operating philosophy that emphasizes the importance of allowing employees to make independent operations-related decisions.

operational aspect of its organization and (b) present them to employees as facts to be accomplished. Instead, an alternative approach occurs when employees are given the "power" to get involved.

Employees can be empowered to make critical decisions concerning themselves and, most importantly, an operation's guests. Many foodservice employees work closely with guests, and numerous problems are more easily resolved when employees are given the power to make it "right" for the guests. In most cases, empowered employees yield a loyal and committed workforce that is more productive, is supportive of management, and will "go the extra mile" for guests. Doing so helps reduce labor-related costs, builds repeat sales, and increases profits.

Employee empowerment also improves interpersonal relationships since the emphasis is on the foodservice operation's team members and guests rather than on its products and services. Other benefits of employee empowerment include an increased opportunity for synergy: the outcome that occurs when several persons working together achieve total results that are greater than if the persons worked independently of each other. Empowerment also places the proper emphasis on a foodservice operator's results rather than on processes, and it increases the satisfaction of guests and employees.

Operators should recognize that some employees resign from a foodservice operation thinking "they didn't quit the company; they quit their manager." Lower and unacceptable levels of respect and appreciation may be a silent reason why some employees leave a foodservice operation or work only to the required minimum performance levels. Empowered team members, however, typically have increased levels of **morale** that, in turn, can lead to reduced absenteeism and low-employee turnover rates.

Several leadership approaches can help maximize opportunities for empowerment within a foodservice operation:

Key Term

Morale: The feelings including outlook, satisfaction, attitude, and confidence that team members have at work.

- ✓ The best leaders know that decision-making authority (power) can be given away because power is not a fixed sum (when an operator has more power the employees have less power). Instead, these operators understand that power is expandable and that trust between leaders and team members is increased when power is shared.
- ✓ Leaders who empower their team members give them the discretion to make decisions. They ask for suggestions and allow staff members to exercise judgment. Can employees respond to unique expectations if they must always follow **standard operating procedures (SOPs)**? The proper response is "they can't," and many excellent ideas are not considered because they have never been stated or considered.

Key Term

Standard operating procedure (SOP): The term used to describe the way something should be done under normal business operating conditions.

✓ Those leading foodservice operations recognize the need to invest in training, coaching, and other professional development activities (see Chapter 8). When this is done, staff members can be prepared for how things should "normally" be done. This, in turn, reduces the need for discretion and maximizes opportunities for SOPs to be followed.

✓ Empowered team members take personal responsibility for their actions. Foodservice operators should allow and encourage them to take ownership of their output as they assume work responsibilities. The concept of **accountability** applies to all team members because team members are accountable to each other and to the foodservice operation.

Key Term

Accountability: The term used to indicate that employees take responsibility for both their performance and the results of that performance.

How can foodservice operators prepare employees for empowerment? First, they should consistently share a core value that leaders help individual team members to better do their jobs. Leaders should also clearly define the boundaries of responsibility that staff members should assume as decisions are made to help move toward attainment of an operation's mission. Operators should explain their expectations and empower employees with important (not just unimportant!) responsibilities.

Employee Team Development

Foodservice operators typically lead several different teams within one operation as they direct the work of their staff. For example, all foodservice employees, whatever their position, represent a team. Within that group, there is a team that typically comes in direct contact with the guests (**front-of-house staff**) and another group of staff members that typically does not (**back-of-house staff**).

It is possible that several persons in the front-of-house are food servers (another team) and yet another team is a group of bartenders. Back-of-house personnel may include a team of cooks and others responsible for food production, and another team of workers responsible for facility clean-up.

Key Term

Front-of-house staff: The employees of a foodservice operation whose duties routinely put them in direct contact with guests.

Key Term

Back-of-house staff: The employees of a foodservice operation whose duties do not routinely put them in direct contact with guests.

Since each team member's work should relate directly to that done by others, all employees must understand the importance of teamwork. Those who cook, for example, may depend upon others to provide clean service ware, and servers require properly prepared menu items if guests are to be consistently pleased when they receive their ordered items.

Employees in different teams must work together to ensure the guests receive the best possible service. If one staff member does not perform required work correctly, negative consequences can occur and create problems ranging from food-borne illness to negative reviews posted online. What one team member does, or fails to do, can impact the work of the entire operation's team, and the success or failure of the operation.

Benefits of Effective Teamwork

Effective teamwork benefits a foodservice operation in several ways. For example, it can increase guest satisfaction and improve productivity because staff cooperate with each other, and there is reduced interpersonal conflict and competition.

Those who have been a member of a properly functioning team know that work is most often more fun when working with others than when working all alone. Teamwork allows more opportunities to celebrate success together and feel encouraged about the work that is accomplished. Each team member may have different tasks and responsibilities, but any progress made toward common goals can be cause for shared celebration.

Effective teams reduce the burnout of their individual members. Burnout is typically the result of prolonged stress that can be caused by overwhelming responsibility. The mental, emotional, and physical exhaustion that can come from working very hard may cause talented individuals to quit their jobs. Effective work teams reduce stress and can improve individual employee retention and performance.

Meaningful and positive workplace relationships are most often the result of effective teamwork. The more team members trust one another to reach common goals, the more opportunities are presented to relate to one another positively and to form close relationships. An improved work environment will exist when employees feel more comfortable working with each other, and being part of an effective team allows them to experience and build those relationships.

Effective and successful teams typically share many key characteristics. The first and, perhaps, most important characteristic is that group members trust each other. This often yields common goals that are defined and accepted by all team members.

Characteristics of Effective Team Leaders

Effective team leaders consistently demonstrate several important traits. First, they have strong interpersonal (people) skills. These leaders allow team members to make decisions and, when appropriate, share important responsibilities with team members who have appropriate knowledge, skills, and experience. They also freely allow team members to establish or, at least, contribute to planning goals for the team. They also solicit input for improving work methods, productivity, and other issues affecting operational goals. These leaders may also serve as the link between the team and other managers.

Effective team leaders practice the "art and science" of leadership to maximize morale levels, and they try to make the best use of the limited resources that are available to them. Finally, effective team leaders inspire active participation in problem-solving and decision-making, and they encourage the presentation of creative work ideas and alternative ways of achieving team goals by all team members.

Cross-functional and Self-directed Work Teams

The use of **cross-functional teams** is one good way to get a group of employees from different work teams together to resolve problems. For example, a concern about slow service for a specific food item ordered by a guest in a table-service operation may be blamed on slow kitchen production by those taking the guest's order. At the same time, food production personnel in the kitchen may blame order takers who fail to mention that there is at least a 15-minute wait for that specific item because of its complexity and extended preparation time. Use of a cross-functional team can likely resolve this problem and generate a preferred method to address the issue.

Key Term

Cross-functional teams: Groups of employees from different operating teams within a foodservice operation who work together to resolve problems.

In some operations, work teams become strong, and the use of **self-directed work teams** is possible. Self-directed work teams are a natural extension of empowerment because they allow affected employees to experience team empowerment.

This type of team requires little on-going direction from a supervisor, and the team is responsible for making many decisions impacting the team. For example, team members may determine the job

Key Term

Self-directed work teams: Groups of employees who combine their talents to work without the influence of traditional manager-based supervision.

assignments for the group's members and/or develop their own work schedules. The hospitality and other service industries have not extensively utilized self-directed teams in the past. In the future, these highly skilled and motivated teams may be utilized to address challenges that are now managed with more traditional procedures as they seek to become the **employer of choice** in their own business communities.

Ethics

It is essential that foodservice operators be ethical. The concept of **ethics** relates to what individuals should and should not do in their interactions with others. It is important to recognize that ethics are not the same as feelings, religion, the law, cultural norms, science, or values. Ethics are important because they provide structure and stabilization for society.

Key Term

Employer of choice: A company whom workers choose to work for when presented with other employment choices. This choice is a conscious decision made when joining a company and when deciding to stay with the employer.

Key Term

Ethics: Standards that help decide between right and wrong and influence behavior toward others.

The headlines in daily news sources regularly reveal stories about the conduct of individuals holding elective offices and top leaders in corporations, and they frequently reveal practices that may be unethical and/or illegal.

Unfortunately, perhaps, there is no standard definition for what is right or wrong behavior in a foodservice setting. Instead, individual foodservice operators, based on their background and personal experience, make their own ethics-related decisions numerous times every day. Many of the most critical issues impacting the foodservice industry relate to ethical concerns. These may include wages paid and benefits granted, hiring and firing decisions, fairness to workers, professional development opportunities, discipline, and other concerns. As a result, the question of what is and is not ethical behavior can sometimes be unclear.

Some reasons why ethical problems can occur include:

✓ Greed
✓ Pressure to meet unrealistic expectations
✓ The belief that "everyone does it"
✓ No organizational "checks and balances" on ethics-related decisions
✓ It is sometimes easy to cheat and not be caught
✓ There may be little punishment for unethical behavior
✓ An organization's reward system encourages success without concern about the ethical tactics used to achieve success

A foodservice operation's ethical culture begins at the top of the organization as operators develop and discuss mission statements and core values with their staff. The concept of proper ethical conduct, like other factors in the mission statement, should be driven down throughout all levels of a foodservice operation.

Effective foodservice operators must be concerned about ethical issues when they make major decisions affecting their organizations and more routine day-to-day operating decisions. Foodservice operators seeking to determine whether a course of action is, or is not, ethical can ask themselves several key questions:

✓ Is it legal?
✓ Does the action negatively affect anyone?
✓ Is the action fair?
✓ Am I being honest?
✓ Can I live with the action without being bothered by it afterward?
✓ Would I be willing to share my actions publicly with others inside and outside of my organization?
✓ Would it be okay if everyone did it?

The leader of a foodservice operation will routinely make numerous decisions with ethical implications. Therefore, it is a good tactic to add the question, "Is it ethical?" along with other factors normally addressed such as "What is best for our guests?", "What will my employees think of me if I do this?", and "How much will it cost or save?"

Ethical conduct should be required of all foodservice operators; however, the difference between right and wrong behavior can sometimes be viewed from different perspectives. For example:

✓ An employee whose responsibility involves purchasing attends a vendors' "Showcase" that offers new products of potential interest to foodservice operators. In addition to special low prices, purchase of these new products provides "points" to purchasers who buy the products that evening. Based on the total "points" given to the purchaser, the employee can select complimentary items such as a smart tablet or expensive sunglasses. Is this acceptable, or should the buyer choose only items that their operation already uses? If acceptance of the gift items is acceptable, who owns them?
✓ A foodservice operator develops a budget for the following year that provides an increase in employee benefits. They then compensate by eliminating some full-time positions and utilizing more part-time employees who do not qualify for these benefits. Is this ethical?
✓ The chef of a foodservice operation is planning a new menu and requests samples from alternative vendors. One vendor's representative delivers the products to the kitchen and provides a special package to the chef with the

comment: "Here are some steaks like we delivered here and you can take home to try on your grill. Let me know what you think." Should the chef be allowed to keep the steaks?

Examples such as these indicate the importance of developing a **code of ethics** designed with input from employees, and it provides needed information about what high-level operators feel is and is not ethical conduct.

The purpose of a code of ethics is to help identify what an organization considers to be acceptable and unacceptable behavior. Action steps that can help establish and emphasize proper ethical conduct include:

Key Term

Code of ethics: A guiding set of principles intended to instruct professionals to act in a manner that is honest and beneficial to all an organization's stakeholders.

✓ Having a mission statement and clearly understanding what the organization stands for.
✓ Being aware of the operation's professional and community image.
✓ Involving everyone who will be affected in the development of the code of ethics.
✓ Implementing organization-wide training and communication as the code of ethics is implemented.
✓ Using specific "ground rules" to define behavior related to issues that are most likely to create problems and ensure they are addressed in the code.
✓ Being ready and willing to enforce all elements of the code fairly and consistently.

Developed correctly, a code of ethics is an excellent tool for identifying how foodservice operators and their employees should interact with, and relate to, all, stakeholders they serve.

Find Out More

The McDonald's corporation is one of the most successful organizations in the history of the foodservice industry. It is unique in many ways, but one area in which it stands out relates to the development and enforcement of its code of ethics.

The code of ethics that McDonald's employees are expected to follow is covered in the corporation's Standards of Business Conduct document. Each year, all employees must sign a document certifying that they've read the information and will consistently follow it.

Since its inception, McDonald's has recognized the relationship between ethical behavior and business success. This was stated clearly by Ray Kroc (1958*), the founder of the McDonald's corporation:

The basis for our entire business is that we are ethical, truthful, and dependable. It takes time to build a reputation. We are not promoters. We are business people with a solid, permanent, constructive ethical program that will be in style . . . years from now even more than it is today.

The McDonald's code of conduct is a document of 40-plus pages and covers major themes including ethics. To review the information in it, enter "McDonald's Standards of Business Conduct" into your favorite search engine and view the results.

*https://corporate.mcdonalds.com/corpmcd/investors/corporate-governance.html#:~: text=In%20short%2C%20what%20Ray%20Kroc,We%20are%20not%20promoters; retrieved October 10, 2023.

As this chapter has indicated, leaders of a foodservice operation are unique individuals charged with a variety of important responsibilities, and leaders may exercise their leadership in many tasks that they feel are best for their organizations. There are, however, laws and regulations that apply to foodservice operations that must be followed by all leaders. Knowing and understanding what these laws are and how they apply to foodservice operations is very important, and they will be the sole topic of the next chapter.

What Would You Do? 1.2

"I can't believe I missed it, . . . and the delivery driver missed it too! But I've counted three times, and we got an extra case of steaks!" said Jonna.

Jonna, the purchasing and receiving agent at the Big Wheel Steakhouse, was talking to Eric, the steakhouse manager.

"So, it was kind of busy when the meat delivery arrived," said Jonna, "the driver and I counted all the boxes of product, and I signed the delivery invoice. But when I started putting the boxes away, I found out we actually had one more case of steaks than was listed on the delivery invoice. So, we got a free box of steaks!"

Assume you were Eric. What would you think about Jonna's comment that your operation received a "free" box of steaks? What message do you think it will send to Jonna and others in the operation with whom she might discuss this event if you decided to keep the steaks? What message do you think it would send to your work team if you decided to notify the meat vendor that you had received, but not been charged for, an extra box of steaks?

Key Terms

Management	Delegate	Front-of-house staff
Asset (business)	Empowerment	Back-of-house staff
Stakeholders	(employee)	Cross-functional teams
Value	Morale	Self-directed work teams
Employee turnover rate	Standard operating	Employer of choice
Mission statement	procedure (SOP)	Ethics
Core values	Accountability	Code of ethics

Operator's 10-Point Tactics for Success Checklist

Evaluate your need for, and the current status of, each of the following operational tactics. For those tactics you think are important, but not yet in place, develop an action plan for its implementation including who will be responsible for the tactic's completion and the target date by which it should be completed.

				If Not Done	
Tactic	Don't Agree (Not Done)	Agree (Done)	Agree (Not Done)	Who Is Responsible?	Target Completion Date
1) Operator understands that the management process involves planning, organizing, directing, controlling, and evaluating how organizational resources are used.	⎯⎯	⎯⎯	⎯⎯		
2) Operator recognizes that both guests and employees are important stakeholders of a foodservice operation, and both have key needs that must be met.	⎯⎯	⎯⎯	⎯⎯		
3) Operator can identify the key differences in four major leadership styles.	⎯⎯	⎯⎯	⎯⎯		
4) Operator recognizes the importance of developing a mission statement and of clearly articulating their organization's core values.	⎯⎯	⎯⎯	⎯⎯		

Tactic	Don't Agree (Not Done)	Agree (Done)	Agree (Not Done)	If Not Done	
				Who Is Responsible?	Target Completion Date
5) Operator understands the need to consistently practice important leadership principles as they direct and lead staff members.	——	——	——		
6) Operator recognizes they are addressing special employee-related concerns as they plan empowerment programs, form employee teams, and address important ethical issues.	——	——	——		
7) Operator realizes the advantages that accrue from developing work teams with optimal levels of employee empowerment.	——	——	——		
8) Operator understands the importance to success of effective employee team development principles.	——	——	——		
9) Operator recognizes the importance to long-term success of ethical behavior in all their dealings with guests and employees.	——	——	——		
10) Operator has developed and utilizes a code of ethics for their organization.	——	——	——		

2

Legal Aspects of Employee Management

What You Will Learn

1) How Employment Law Affects Foodservice Operators?
2) The Impact of Federal Law on Employee Management
3) The Impact of State Law on Employee Management
4) The Impact of Local Laws and Ordinances on Employee Management

---

Operator's Brief

In this chapter, you will learn that, as a foodservice operator, you have significant freedom in how you manage your employees and your business. That freedom is not, however, unlimited.

All foodservice operators in the United States must follow applicable federal, state, and local laws and ordinances. There are numerous important federal laws that will directly impact employee management and some of the most important are addressed in the chapter and include:

✓ Clayton Act
✓ Railway Labor Act (RLA)
✓ National Labor Relations Act (Wagner Act)
✓ Fair Labor Standards Act (FLSA)
✓ Equal Pay Act
✓ Civil Rights Act (Title VII)
✓ Age Discrimination in Employment Act
✓ Pregnancy Discrimination Act
✓ Americans with Disabilities Act (ADA)
✓ Family and Medical Leave Act (FMLA)
✓ Patient Protection and Affordable Care Act (ACA)

In addition, individual states can enact their own employment-related laws and many of these can affect what operators can and cannot do regarding the management of their work teams. Finally, you will learn that local governments including counties, municipalities, and cities are also free to enact laws and ordinances that impact employee management, and these are addressed because they can affect how one can legally manage employees and your foodservice operation.

CHAPTER OUTLINE

Understanding Employment Law
Federal Laws Affecting Employee Management
    Pre-1964 Federal Employment Legislation
    Post-1964 Federal Employment Legislation
State Laws Affecting Employee Management
Local Laws and Ordinances Affecting Employee Management

Understanding Employment Law

The foodservice industry is highly regulated. Experienced operators know they must follow a large number of laws and regulations in regard to, for example, proper labeling of the menu items they sell, maintaining their physical facilities in accordance with local health code ordinances, and reporting accurate sales and tax collections to the proper taxing authorities. There are also numerous key laws and regulations regarding employees, and, as a result, every foodservice operator must understand the impact of **employment law**.

Key Term

Employment law: The body of laws, administrative rulings, and precedents that address the legal rights of workers and their employers.

Employment law in the United States arose in many cases because of workers' demands for better working conditions. Wherever and whenever these workers' demands were deemed reasonable by most of society and/or the courts, legislation was enacted and became part of the country's accepted employment practices.

Today, employment laws are still proposed by various segments of society to ensure fairness in the workplace. However, the societal view of what constitutes fairness can often be controversial and ever-changing. For example, some employment-related legislation that would have been considered quite radical in the 1800s is today commonly accepted. Consider that the now-accepted concept that the rights of female employees should be equal to those accorded to men was certainly not the norm in the 1800s. In fact, it was 1920 before women in the United States gained the right to vote!

Citizens of individual countries, states, or cities may vary greatly in their own views of what constitutes fairness in employment. Not surprisingly, various types of employment laws may be enacted in different **jurisdictions** and foodservice operators must be keenly aware of individual employment laws that directly affect them, their operations, and their employees.

Key Term

Jurisdiction: The geographic area within which a court or government agency may legally exercise its power.

Foodservice operators must understand that the wrong employee-related decision can subject their companies (and themselves) to significant legal liability. Multi-million dollar jury awards levied to penalize companies found guilty of improper employment practices are common in the United States. As a result, it is critical that operators recognize and follow all employment-related laws that directly affect them and their employees.

Sometimes the laws directly related to employment in a foodservice operation are general (e.g., federal laws in the United Sates relating to the rights of workers to unionize). At other times, laws related to employment may best be understood in the context of a particular segment of employee management. For this reason, laws related specifically to employee health and safety (see Chapter 4), employee recruiting and hiring (see Chapters 5 and 6), and compensation (see Chapter 10) are closely examined in the applicable chapters.

Of course, foodservice operators are not expected to be attorneys. A lack of understanding about employment laws, however, can easily produce problems that might require the services of a qualified (and often expensive!) attorney.

Experienced operators know that lawsuits and litigation are expensive and time consuming. Most would also agree that the negative publicity associated with highly publicized lawsuits can be a real detriment to their business and even their careers. For these reasons, all foodservice operators should take great care to ensure that their actions do not inadvertently create troublesome legal issues for their employees, themselves, and their companies.

Foodservice operators in the United States have many partners in their employee-related activities and decision-making because the foodservice industry is regulated by a variety of federal, state, and local governmental entities. Operators interact with governmental entities in numerous ways, and they must observe the procedures and regulations established by these agencies. Operators must fill out forms and paperwork, obtain operating licenses, and maintain their property according to specified codes and standards. They must also provide a safe working environment, and, when required, open their facilities for periodic inspections.

It should come as no surprise that a society, working through its governmental structures, implements and often revises its rules of employee–employer conduct and responsibilities. As an example, foodservice operators in the 1950s would not likely find a law requiring a specific number of automobile parking spaces to be designated for their disabled workers and guests.

In the 1950s, society did not deem it essential to grant special parking privileges to those who were disabled. Now the **Americans with Disabilities Act (ADA)**, a law created in 1990 by the federal government, requires an operation to provide a specific number of designated "handicapped" parking spaces. *Note*: These spaces are generally located close to the main entrances of buildings to ensure easy access.

Key Term

Americans with Disabilities Act (ADA): A federal law prohibiting discrimination against persons with disabilities.

The ADA is addressed here to illustrate that laws evolve just as society evolves. Knowing current law is important but understanding that an operator must keep abreast of changes in the law to ensure that the facilities they operate are managed legally is just as important.

Just as the federal government has played (and will continue to do so) an important regulatory role in the foodservice industry, the various state governments also serve both complementary and distinct regulatory roles. These are important and some roles are complementary in that they support and amplify efforts undertaken at the federal level. However, they can also be distinct in that they regulate some areas for which they have sole responsibility.

Recognizing that state and even local employment laws and regulations may affect the actions of foodservice operators more often than federal regulations is important. Codes and ordinances established at the state or local level can often be strict, and they should be carefully enforced. The penalties for violating these laws can be as severe as those imposed at the federal level.

In general, each state regulates significant parts of the employee–employer relationship occurring within its borders. For example, items such as worker-related **unemployment compensation benefits** to be paid, worker safety issues, and at-work injury compensation generally fall to the state entity charged with regulating the workplace.

Key Term

Unemployment compensation benefits: Benefits paid to employees who involuntarily lose their employment without just cause.

Also, in most communities an agency of the court system (sometimes called a "friend" of the court) has the responsibility of assisting creditors in securing payment for legally owed debts. These can include a variety of court-ordered payments such as for

Location	Jurisdiction	Minimum Wage	% of Federal Minimum Wage
United States	Federal	$7.25	100%
State of Washington	State	$15.74	217%
Seattle, WA	Local	$18.69	258%

Figure 2.1 Minimum Wage Comparison

child support. In such cases, a foodservice operator can be ordered by the court to **garnish** an employee's wages.

In some cases, the decisions of local governments can have a very significant impact on a foodservice operation's profitability. To cite just one example, at the time of this book's publication, the **minimum wage** in Seattle, WA, was $18.69 an hour. At the same time, the minimum wage in Washington state was $15.74 and the federal minimum wage (unchanged since 2009) was $7.25 per hour.

Using the Seattle minimum wage law as an example, a review of Figure 2.1 illustrates the large impact that local jurisdictions can have on, in this instance, a foodservice operation's cost of labor.

Key Term

Garnish(ment): A court-ordered method of debt collection in which a portion of a worker's income is paid directly to one or more of that worker's creditors.

Key Term

Minimum wage: The lowest wage per hour that a worker may be paid as mandated by law.

Employment laws in the United States may be enacted at the federal, state, or local level. They reflect the desires of citizens and, ultimately, their elected officials and courts. Most foodservice professionals agree that all workers are best protected when employers, employees, and governmental entities work together to protect wages, benefits, safety, and health in a manner that allows their employers to operate profitably.

Federal Laws Affecting Employee Management

Federal law impacts many areas in which foodservice operators legally manage their employees. While there are several ways to examine the employment-related legislation passed by the federal government, one good way is to review significant employment legislation enacted before and after passage of the landmark Civil Rights Act of 1964.

Pre-1964 Federal Employment Legislation

In the view of many U.S. history scholars, the most significant employment legislation passed before 1964 related to unionized workers and took place during the 1930s. However, several important pieces of labor-related legislation were passed in the very early 1900s. One of the most noteworthy was the Clayton Act of 1914, which legitimized and protected workers' rights to join a **labor union**.

In 1926, Congress passed the Railway Labor Act that required employers to bargain collectively, and it also prohibited discrimination against unions. It applied originally to interstate railroads but, in 1936 it was amended to include airlines engaged in **interstate commerce**.

Key Term

Labor union: An organization that acts on behalf of its members to negotiate with management about the wages, hours, and other terms and conditions of the membership's employment.

Key Term

Interstate commerce: Commercial trading or the transportation of people or property between or among states.

The rights of employees to unionize were established with the passage of the National Labor Relations Act (NLRA) of 1935, more popularly known as the Wagner Act. The NLRA applied to all firms and employees engaging in activities affecting interstate commerce. This law's impact included coverage of restaurant workers. They and most other workers were guaranteed the right to organize and join labor movements, to select representatives and bargain collectively, and, to strike. It also expressly prohibited employers from:

✓ interfering with the formation of a union
✓ restraining employees from exercising their right to join a union
✓ imposing any special conditions on employment that would discourage union membership
✓ discharging or discriminating against employees who reported unfair labor practice
✓ refusing to bargain in good faith with legitimate union leadership

Historically, the great majority of foodservice workers have not been unionized, and this is especially so in smaller foodservice operations. Operators leading employees in organizations that are unionized, however, must understand the unique employer/employee relationship that exists in a unionized workplace and especially how it relates to employee termination (see Chapter 11).

In 1938 Congress passed the Fair Labor Standards Act (FLSA). The main objective of this act was to eliminate labor conditions deemed "detrimental to the maintenance of the minimum standards of living necessary for health, efficiency, and well-being of workers."[1]

---

1 www.dol.gov/sites/dolgov/files/WHD/legacy/files/FairLaborStandAct.pdf. Retrieved March 20, 2023.

This law requires employers to pay overtime for hours worked in excess of 40 per week (defined as 7 consecutive 24-hour periods).

The FLSA also prohibits **child labor** in all industries engaged in producing goods for sale in interstate commerce. The act sets the minimum age at 14 for employment outside of school hours in nonmanufacturing jobs, at 16 for employment during school hours, and at 18 for hazardous occupations.

In 1963 Congress passed the Equal Pay Act. This law prohibits employers from paying women and men different wages when the work performed requires equal skill, effort, and responsibility and when it is performed under similar working conditions.

Key Term

Child labor: Work that deprives children of their childhood, their potential, and their dignity, and that is harmful to physical and mental development.

Find Out More

For many foodservice operators, tip credits are an important employment-related issue. Essentially tip credits allow employers to include tips as part of their workers' wage calculations.

Foodservice operators can credit a portion of an employee's received tips *toward* the required minimum wage pay so the full amount of tip credits does not need to be paid in the full amount. In other words, tip credits are not a deduction from pay but, instead they are included in the minimum wage payment calculation.

Under FLSA rules, an employer that elects to take a tip credit against the Federal minimum wage must pay the tipped employee a direct cash wage of at least $2.13 per hour (at the time this book is published). Provided that the employer meets certain requirements, the employer may then take a credit against its wage obligation for the difference, up to $5.12 per hour if the employees' tips are sufficient to fulfill the remainder of the minimum wage.

However, regardless of the FLSA rules, currently there are eight states and one territory that don't allow a tip credit in the United States. The states are California, Oregon, Washington, Montana, Nevada, Minnesota, Nevada, Alaska, and Guam is the territory.

Even within the states that do allow tip credits significant variations exist. To learn more about the requirements for legally taking the tip credit in a specific state, enter "tip credit requirements in (state name)" and review the results.

Post-1964 Federal Employment Legislation

Perhaps the single most significant piece of federal legislation affecting the workplace was passed in the aftermath of a true American tragedy. The assassination of John F. Kennedy in November 1963, resulted in the Lyndon Baines Johnson presidency. On November 27, 1963, addressing Congress and the nation for the first time as president, Johnson called for passage of a sweeping civil rights bill as a monument to the fallen Kennedy. "Let us continue," he said, promising that "the ideas and the ideals which [Kennedy] so nobly represented must and will be translated into effective action."[2]

In June 1964, Congress passed the Civil Rights Act of 1964. It was the most important piece of civil rights legislation in the nation's history, and on July 2, 1964, President Johnson signed it into law.

The Civil Rights Act of 1964 contains several important sections, but for the foodservice industry, the most important of these is **Title VII**.

Title VII of the Civil Rights Act of 1964 outlawed discrimination in employment in all businesses on the basis of race, color, religion, sex, or national origin. In 1972, the passage of the Equal Employment Opportunity Act, a modification of the Civil Rights Act of 1964, resulted in the formation of the **Equal Employment Opportunity Commission (EEOC)**.

The EEOC enforces the antidiscrimination provisions of Title VII. The EEOC investigates, mediates, and sometimes even files lawsuits on behalf of employees. Businesses that are found to have discriminated against one or more employees can be ordered to compensate the employee(s) for damages in the form of lost wages, attorney fees, and other expenses.

Title VII also provides that individuals can sue their employers on their own. In most cases, an individual must file a complaint of discrimination with the EEOC within 180 days of learning of the discrimination or lose the right to file a lawsuit.

In the late 1970s, U.S. courts began holding that **sexual harassment** is sex discrimination prohibited under the Civil Rights Act, and in

Key Term

Title VII: A federal employment law that prohibits employment discrimination based on race, color, religion, sex, or national origin.

Key Term

Equal Employment Opportunity Commission (EEOC): The entity within the federal government assigned to enforce the provisions of Title VII of the Civil Rights Act of 1964.

Key Term

Sexual harassment: Unwelcome sexual advances, requests for sexual favors, and other verbal or physical conduct of a sexual nature.

---

2 https://www.presidency.ucsb.edu/documents/address-before-joint-session-the-congress-0. Retrieved October 23, 2023.

1986 the U.S. Supreme Court held that sexual harassment is sex discrimination and thus is prohibited by Title VII.

Over time, Title VII has been supplemented with legislation prohibiting discrimination based on pregnancy, age, disability, and other worker characteristics. In 2020, the U.S. Supreme Court issued a landmark decision holding that Title VII prohibits discrimination against employees based upon sexual orientation and transgender status.

In the 6-3 Opinion of the Court, written by Justice Gorsuch, a majority of justices held that a "straightforward" rule emerges from the ordinary meaning and application of Title VII's prohibition against sex discrimination:

> [F]or an employer to discriminate against employees for being homosexual or transgender, the employer must intentionally discriminate against individual men and women in part because of sex. That has always been prohibited by Title VII's plain terms—and that should be the end of the analysis.[3]

As a result of the court's 2020 ruling, it is now unlawful to discriminate in employment because of sexual orientation or gender identity.

Technology at Work

Universal changes in employment-related law can happen quickly and have the impact of affecting employers immediately.

One good example of this occurred in 2020, when the Supreme Court of the United States issued its landmark decision in the case *Bostock v. Clayton County*, which held that the prohibition against sex discrimination in Title VII of the Civil Rights Act of 1964 (Title VII) includes employment discrimination against an individual based on sexual orientation or transgender status.

Title VII protects job applicants, current employees (including full-time, part-time, seasonal, and temporary employees), and former employees if their employer has 15 or more employees. *Note*: Employers with fewer than 15 total employees are not covered by Title VII.

With the exception of the "note" in the previous sentence (number of employees), Title VII protects employees regardless of citizenship or immigration status, in every state, the District of Columbia, and the United States territories.

The Internet provides foodservice operators with an effective tool for monitoring changes in employment law. It is a good idea for foodservice operators to regularly check for changes in employment law that will directly affect them. You can do so by regularly entering "Recent legislation affecting restaurant operators" in your favorite search engine and viewing the results.

---

3 https://workforce.com/news/the-top-7-recent-employment-law-cases-you-should-know. Retrieved March 26, 2023.

The Civil Rights Act of 1964 also makes it illegal for employers to discriminate when hiring and setting the terms and conditions of employment. Labor unions are also prohibited from basing membership or union classifications on race, color, religion, sex, or national origin. The law also prohibits employers retaliating against employees or potential employees who file charges of discrimination against them, refuse to comply with a discriminatory policy or who participate in an investigation of discrimination charges against the employer.

In some very few cases, Title VII permits a **bona fide occupational qualification (BFOQ)** to be used to legally discriminate among workers. It is important to understand, however, that the government, through the EEOC, and not an individual foodservice operator, can authorize the legal use of a BFOQ.

Thus, for example, even if an Asian foodservice operator wanted to hire only Asian servers, or a Mexican foodservice operator wanted to hire only Spanish speaking Hispanics as servers, neither of those race-based decision situations would be permitted as a BFOQ, under Title VII.

In addition to the Civil Rights Act of 1964, many states also have their own civil rights laws that prohibit discrimination. Sometimes state laws are more inclusive than the Civil Rights Act in that they expand protection to workers or employment candidates in categories not covered under the federal law (such as age, marital status, sexual orientation, and certain types of physical or mental disabilities).

Key Term

Bona fide occupational qualification (BFOQ): A specific job requirement for a particular position that is reasonably necessary to the normal operation of a business, and therefore allows discrimination against a protected class (for example, choosing a female model when photographing an advertisement for lipstick).

State civil rights laws may also have stricter penalties for violations, including fines and/or imprisonment. It is important to remember that foodservice operators must be aware of all civil rights laws in effect in the location where they are working. While a state or local law is not permitted to take away employee rights granted at the federal level, they are allowed to add to them. Thus, as in this example, an employee group not protected by federal law may be granted protection by a more favorable (to the employee) state or local law.

The Age Discrimination in Employment Act (ADEA) of 1967 was initially passed to prevent the widespread practice (at that time) of requiring employees to retire at age 65. As workers' life spans increased, society felt it made little sense for employees to be forced to retire when many could and wanted to remain on the job.

The ADEA originally gave protected-group status to those workers between the ages of 39 and 65. Since 1967, the act has been amended twice: once in 1978, when the mandatory retirement age was raised to 70, and then again in 1986, when the mandatory retirement age was removed altogether.

The ADEA makes several factors including age preferences, limitations, or specifications in job notices or advertisements unlawful. As a narrow exception to that general rule, job notices or advertisements may specify an age limit in the rare circumstances where age is shown to be a BFOQs reasonably necessary to operate the business. For example, airline pilots may still be required to retire at a certain age because of evidence that increasing age causes a decline in piloting abilities.

Customer preference is not a rationale that will result in the granting of an age-related BFOQ by the EEOC. For example, a foodservice operator who decides that their customers like young (and attractive) female servers better than older ones would not be allowed to hire only younger servers. They would also not be allowed to terminate servers as they became older, even if the bar owner believed doing so would be good for business.

The ADEA does not specifically prohibit an employer from asking an applicant's age or date of birth. In the foodservice industry, for example, establishing an applicant's age before hiring him or her to serve alcoholic beverages or to operate certain types of potentially dangerous kitchen equipment may be legally necessary. *Note*: Operators' inquiries during the employee selection process may deter older workers from applying for employment or may otherwise indicate possible intent to discriminate based on age. Therefore, requests for age information are closely scrutinized by the EEOC to ensure inquiries are made for a lawful rather than prohibited purpose.

Under the ADEA, it is unlawful to discriminate against a person because of their age with respect to any term, condition, or privilege of employment, including hiring, firing, promotion, layoff, compensation, benefits, job assignments, and training. Harassing an older worker because of age is also prohibited.[4]

The Older Workers Benefit Protection Act of 1990 amended the ADEA to specifically prohibit employers from denying benefits to older employees. This law was passed to prevent reducing employment benefits such as medical insurance based on an employee's age.

The Pregnancy Discrimination Act of 1978 is an amendment to Title VII of the Civil Rights Act of 1964. The Pregnancy Discrimination Act made discrimination based on pregnancy, childbirth, or related medical conditions unlawful sex discrimination under Title VII.

The law requires that women affected by pregnancy or related medical conditions must be treated in the same way as other applicants or employees with similar abilities or limitations. An employer cannot refuse to hire a woman because of

---

4 www.eeoc.gov/laws/guidance/fact-sheet-age-discrimination. Retrieved March 29, 2023.

her pregnancy-related condition as long as she is able to perform the major functions of her job. Also, an employer cannot refuse to hire her because of prejudices against pregnant workers or the prejudices of co-workers, clients, or customers.

Pregnant employees must be permitted to work as long as they can perform their jobs and must be allowed to take a "reasonable" amount of unpaid leave after a baby's birth. Although the law does not specifically establish a definition for "reasonable" time off after a baby's birth, many organizations consider six weeks to be reasonable, and that time frame has been supported by the EEOC.

What Would You Do? 2.1

"I don't get it," said Tonya Zollars, the owner of Zollar's Steakhouse. "I think she's great!"

Tonya was talking about Naomi Yip, the restaurant's assistant sous chef. Naomi had been with the restaurant for three years and Tonya considered her to be one of the best and brightest culinary artists ever to work at Zollars.

The restaurant's sous chef's position was now open due to a retirement, however, Thomas Hanks, the restaurant's Executive Chef was not recommending Naomi fill the sous chef's vacancy.

"Well," said Thomas, "You know Naomi's gonna have a baby. She's due in five months. I like Naomi a lot, but you know that I count on the sous chef to fill in for me when I'm gone and to basically manage all of our pre-prep. She told me she's excited to have her first child and is looking forward to taking six weeks off, then spending as much time as possible with her baby. And you know as well as I do that at the next level, 50-plus hours weekly for the sous chef's job is the norm around here. You can't do that when you have a new baby."

"And that's why you want someone else to fill the position?" asked Tonya.

"That's right," said Thomas, "and I think we should post an online recruiting ad for our vacancy right away!"

Assume you were Tonya. What would think you about Thomas' comment that "You can't do that when you have a new baby" and his rationale for bypassing Naomi for the sous chef promotion? What legal issues do you think may be in play in this scenario? What ethical issues may be in play? Would you promote Naomi?

The Americans with Disabilities Act (ADA) was enacted in July 1990, and it prohibits discrimination against people with disabilities. The ADA is a five-part piece of legislation, but Title I of the Act focuses primarily on employment.

Three different groups of individuals are protected under the ADA:

1) An individual with a physical or mental impairment that substantially limits a major life activity. Some examples of what constitutes a "major life activity"

under the act are seeing, hearing, talking, walking, reading, learning, breathing, taking care of oneself, lifting, sitting, and standing.
2) A person who has a record or history of a disability.
3) A person who is "regarded as" having a disability.

Employers cannot reduce an employee's pay simply because they are disabled. Likewise, they cannot refuse to hire a disabled candidate if, with **reasonable accommodation**, it is possible for the candidate to perform the job.

Not surprisingly, one of the significant issues regarding ADA is the practical application of the word *reasonable* when considering how to best accommodate a disabled job candidate or employee. U.S. courts have held that restructuring a job to shift a minor (nonessential) responsibility for a task from a disabled to a nondisabled employee is a reasonable accommodation.

Key Term

Reasonable accommodation: A change to the application or hiring process, to the job, to the way the job is done, or the work environment that allows a person with a disability who is qualified for the job to perform the essential functions of that job and enjoy equal employment opportunities.

In addition, allowing disabled workers extra unpaid leave when it does not present a hardship to the business is considered a reasonable accommodation. *Note*: An employer is not, however, required to provide a disabled worker with more paid leave than the employer provides its nondisabled workers.

Certain things are not considered reasonable accommodations and are therefore not required. For example, an employer does not have to eliminate a primary job responsibility to accommodate a disabled employee. In addition, an employer is not required to lower productivity standards that apply to all employees. *Note*: The employer may be required to provide reasonable accommodations to enable an employee with a disability to meet the productivity standard.

An employee's request for reasonable accommodation or the granting of a reasonable accommodation is typically a matter to be discussed only between the employer and the employee. An employer may not disclose that a disabled employee is receiving a reasonable accommodation because this usually amounts to a disclosure that the individual has a disability. (The ADA specifically prohibits the disclosure of an employee's medical information except in very limited situations, which never includes disclosure to co-workers.)

Even with the passage of the ADA, an employer does not have to hire a disabled applicant who is not qualified to do a job. The employer can still select the most qualified candidate, provided that no applicant was eliminated from consideration because of a qualified disability.

Although the law in this area can and does change, Figure 2.2 lists conditions that currently meet the criteria for a qualified disability and are protected under the ADA.

AIDS, HIV, and its symptoms	Cerebral palsy	Migraine headaches	Thyroid gland disorders
Asthma	Diabetes	Muscular dystrophy	Loss of body parts
Blindness or other visual impairments	Epilepsy	Paralysis	
Cancer	Heart disease	Complications from pregnancy	

Figure 2.2 Physical Conditions Protected Under the ADA

Find Out More

The Americans with Disabilities Act (ADA) requires that employers, including foodservice operators, make reasonable accommodations when possible to ensure fairness in the hiring of workers who have a protected physical condition. Sometimes, however, what amounts to "reasonable accommodation" is not easily defined.

Currently, the EEOC's Enforcement Guidance cites the following as possible reasonable accommodations:

✓ Making existing facilities accessible
✓ Job restructuring
✓ Part-time or modified work schedules
✓ Acquiring or modifying equipment
✓ Changing tests, training materials, or policies
✓ Providing qualified readers or interpreters
✓ Reassignment to a vacant position

It is also important to recognize that some foodservice operators are exempted from ADA requirements. Businesses with fewer than 15 employees are not covered by the employment provisions of the ADA. Moreover, a covered employer does not have to provide reasonable accommodation that would cause an "undue hardship" (see Chapter 7).

Undue hardship is defined as an action requiring significant difficulty or expense when considered in concert with factors such as an organization's size, financial resources, and the nature and structure of its operation.

To find out more about the ADA-related requirements affecting foodservice operators, enter "ADA requirements for restaurant employers" into your favorite search engine and view the results.

The Family and Medical Leave Act (FMLA) was enacted in 1993. This law allows an employee to take unpaid leave because of pregnancy or to care of a sick family member. The FMLA currently requires employers of 50 or more employees (and all public agencies) to provide workers up to 12 weeks of unpaid, job-protected leave in a 12-month period:

✓ for the birth and care of a child
✓ for placement with the employee of a child for adoption or foster care
✓ to care for a serious illness of the employee or an immediate family member
✓ for a serious health condition that makes the employee unable to perform the essential functions of their job

Under the FMLA, employers can require a request for leave be certified as necessary by the healthcare provider of the eligible employee or of the child, spouse, or parent of the employee, as appropriate. When the employer requests it, the employee is required to provide, in a timely manner, a copy of the certification to the employer.

In 2010, the U.S. Congress passed the Patient Protection and Affordable Care Act (ACA). This law is especially important for foodservice operators for two reasons: (1) the actual content and impact of the law and (2) the history of the ACA provides operators with a recent (and continuing) example of how employment laws in the United States are proposed, debated in the public arena and, in many cases, enacted in one form or another.

The ACA contains a variety of key provisions that directly affect workers and companies in the hospitality industry including:

✓ Guaranteeing the ability of all individuals to obtain healthcare coverage regardless of their pre-existing health conditions
✓ Providing health insurance cost subsidies for low-income individuals and families
✓ Banning annual and lifetime caps on insurance coverage for all individuals
✓ Providing tax incentives for small businesses that provide health coverage for their workers and pay at least 50% of the cost of the coverage
✓ Fining affected employers who do not provide adequate health coverage to their full-time employees (those who work over 30 hours per week or 130 hours per month)

Enforcement of the parts of the ACA related to employers and their compliance with it is assigned to the Internal Revenue Service (IRS). Regardless of an operator's personal position about the value of the ACA, the dialogue on the law and other issues related to healthcare and the public interest will no doubt continue. The outcomes of these dialogues will ultimately shape changes in existing Federal employment laws (summarized in Figure 2.3), and these will directly affect what foodservice operators can or must do as they legally operate their businesses.

Effect of the Legislation	Law	Enacted
Protected workers' right to join labor unions	Clayton Act	1914
Prohibited employment discrimination against union members	Railway Labor Act (RLA)	1926
Gave unionized workers the right to collective bargaining and to strike	National Labor Relations Act (Wagner Act)	1935
Established overtime pay requirements and child labor laws	Fair Labor Standards Act (FLSA)	1938
Mandated equal pay to men and women for equal jobs	Equal Pay Act	1963
Prohibited employment discrimination based on race, color, religion, sex, or national origin; later revisions (in 1972) established the EEOC	Civil Rights Act (Title VII)	1964
Prohibited employment discrimination based on age	Age Discrimination in Employment Act	1967
Prohibited employment discrimination based on pregnancy	The Pregnancy Discrimination Act	1978
Prohibited discrimination based on workers' disabilities	Americans with Disabilities Act (ADA)	1990
Allowed affected workers to receive unpaid leave for selected medical and family health-related issues	Family and Medical Leave Act (FMLA)	1993
Addressed healthcare insurance coverage for all U.S. citizens	Patient Protection and Affordable Care Act (ACA)	2010

Figure 2.3   Selected Labor Legislation Enacted by the U.S. Federal Government

State Laws Affecting Employee Management

Individual states have much freedom if they desire to establish their own laws related to employees and employee management. As previously illustrated in Figure 2.1, for example, many states have established their own minimum wage laws. In those states, employees are covered by the law that is most favorable to them (the state or federal wage) that provides the highest employee compensation.

The differences in state employment laws can be significant, and foodservice operators must be aware of those that relate to the state(s) in which they do business. To further illustrate this fact, consider the very specific differences contained in the work-related laws of selected states detailed in Figure 2.4.

ARKANSAS: Employers of workers who receive board, lodging, apparel, or other items as part of the worker's employment may be entitled to an allowance for such board, lodging, apparel, or other items, not to exceed 30 cents per hour, credited against the minimum wage.

MICHIGAN: Workers younger than age 18 are entitled to a 30-minute meal break after 5 hours of work. Michigan law does not require a meal break for workers older than age 18.

NEW HAMPSHIRE: An employer cannot require an employee to work more than 5 hours without a 30-minute meal break. An employee who reports to work at the employer's request is entitled to be paid a minimum of 2 hours' wages.

RHODE ISLAND: Time and one-half premium pay for work on Sundays and holidays in retail and certain other businesses is required under two laws that are separate from the minimum wage law.

VERMONT: State minimum wage is increased annually by law.

WASHINGTON: No employer may employ a minor without a work permit from the state along with permission from the minor's parent or guardian and school.

Figure 2.4 Selected State-Enacted Employment Legislation

Clearly, individual states have much latitude as they enact their own employment-related laws, and that latitude is regularly exercised.

Foodservice operators must also recognize that while states have a great deal of freedom to enact their own worker related laws, even on an identical issue various states may have enacted diverse legislation. To cite just one example of this, all states have regulations addressing the age at which a worker may serve as a bartender. However, as shown in the sampling of states included in Figure 2.5, different states have placed different restrictions on the allowable age to serve as a bartender.

State	Age Requirement for Bartending
Arkansas, Colorado, Michigan, New York, Texas, West Virginia	18
Arizona, Idaho, Nebraska, New Mexico, North Dakota	19
Kentucky	20
Alabama, California, Illinois, Mississippi, Nevada, Washington	21

Figure 2.5 Selected States Age to Serve as Bartender

> **Technology at Work**
>
> In many cases, foodservice operators will not know some of the applicable employment laws of a state until they work there. Fortunately, the Internet provides operators with an easy tool for identifying key pieces of state legislation important to the management of their businesses.
>
> To cite just one example, various states have different requirements addressing the issuing of work permits for minor workers.
>
> To practice using the Internet to find specific employment-related information such as work permits in a state, enter "work permits requirements for (name of state)" in your favorite search engine and review the results.

Local Laws and Ordinances Affecting Employee Management

Counties, municipalities, and cities have a great deal of latitude in enacting legislation that directly affects foodservice operators. Examples include rules and regulations regarding when it is allowable to serve alcoholic beverages, smoking-related prohibitions, and rules regarding the number of handicapped parking spaces to be offered.

Local laws can directly affect how employees are managed, so it is important to recognize the potential impact of local employment-related laws and **ordinances**. For example, it is not generally illegal to discriminate against prospective or current employees based on their appearance. The reason: Appearance is *not* listed in the Civil Rights Act as a **protected class**.

Key Term

Ordinance: A piece of legislation enacted by a municipal authority.

Key Term

Protected class: A group of people who have special legal protection against discrimination in the workplace based on specifically identified traits.

In fact, there are employers in the foodservice industry (and other industries as well!) that often place a premium on attractive workers (especially front-of-house employees including servers and bartenders who interact with guests).

Of course, if what an employer finds "unattractive" about a person is their age, race, ethnicity, disability, or other protected class characteristic addressed by federal law, that employer would be in violation of Title VII. For example, if a foodservice operator decided that only people who were "White" or "Asian" or "under 30" were attractive, it would likely be very easy to prove a Title VII violation even if employers "claimed" they were only discriminating based on attractiveness.

The ability of a foodservice operation to discriminate based on attractiveness is not universal in the United States. Some municipalities including San Francisco and Santa Cruz (both in California) include "physical appearance" or "physical attractiveness" in local ordinances as a protected class (similar to race, color, religion, sex, and national origin) so workers in these jurisdictions cannot be discriminated against on the basis of their appearance.

As a second example of the potentially significant impact of local laws and ordinances, **living-wage** mandates have been adopted that may impact foodservice operators in some jurisdictions.

The first living-wage law was passed in Baltimore in 1994. The local ordinance passed in that city stipulated that businesses holding service contracts with the city pay a minimum of $6.10 per hour, increasing to $7.70 as of July 1998, and thereafter moving in step with inflation. A single mother working full time at $7.70 per hour would (at that time) have been able to live with her child above the federally defined poverty line.

Within four years of the Baltimore ordinance, living-wage laws passed in New York, Los Angeles, Chicago, Boston, Milwaukee, Jersey City, Durham, Portland, Oregon, and at least eight other cities. Today, more than 160 cities and counties have enacted such measures although the definition of which employers are mandated to pay a living wage and how that amount is established varies greatly.

For foodservice employers, violations of an employment law, regardless of the entity that has enacted it, can create adverse publicity, the loss of significant time and money to address legal issues, and, sometimes, even the closure of their businesses.

Subsequent chapters of this book will, as appropriate, address additional laws that affect foodservice operators. However, as society's view of what constitutes a desirable workplace continues to evolve, legislation affecting employment will also likely continue to evolve at the federal, state, and local levels. All levels of society, through their governments, will adopt regulations that they believe are in the best interests of the communities they represent.

Therefore, professional foodservice owners and operators must stay informed about pending legislation and consider actively taking part in public opinion debates that help shape governmental policies. The reason: problems might

Key Term

Living wage: The minimum hourly wage necessary for a person to achieve a subjectively defined standard of living. In the context of developed countries such as the United States, this standard generally expresses that a person working 40 hours per week with no additional income should be able to afford a specified quality or quantity of housing, food, utilities, transportation, and healthcare.

otherwise arise when those without an understanding of the foodservice industry propose legislation with excessive costs and/or infringement upon individual rights that exceed the societal value of implementing proposed regulations. This is a reason (obtaining information and making their voices heard) that many food-service operators become active in one or more professional trade associations at the local, state, or federal levels.

The leaders of a foodservice operation are unique individuals charged with a variety of important tasks. In addition to monitoring the legal environment of employee management, foodservice operators must also provide a workplace that is safe and healthy for their employees and guests. Ensuring the health and safety of work teams and guests is an essential, and it will be the sole topic of the next chapter.

What Would You Do? 2.2

"I'm just saying, I don't think you should schedule Ruth on Friday night. We're going to be too busy!" said Betty Stout, the dining room supervisor at the Reef Seafood House.

Betty was talking to Ray Swann, the Reef's general manager and the individual in charge of establishing the weekly work schedules for front and back-of-house employees. Ray was preparing the employee schedule for the next week, and they were discussing Friday night, historically the Reef's busiest meal period of the week.

"What do you mean?" replied Ray, "I think Ruth has been with the Reef longer than just about anybody else here (more than 30 years). The guests love her, and no one knows more about our operation than she does."

"That's just my point," replied Betty, "She knows a lot, but when it gets busy, she has trouble keeping up. She's not as fast as the younger servers anymore, and she can't carry as much weight on a tray. That means extra trips to the kitchen, and slower service times for our guests. When Ruth is working the other servers must pull extra weight. And that's just not fair to them. Let's put Ruth on the Tuesday night schedule instead of Friday night."

"I don't know," said Ray "Don't all our servers have limitations in some way or other? They're not all identical. Besides, our volume is less on Tuesday, the tables are smaller, and so are the tips. I'm not sure Ruth will like that!"

Assume you were Ray. Do you think that Betty is suggesting you schedule Ruth differently than other employees because of her age or her performance? Do you think that Betty's arguments would hold up in the face of a potential legal challenge brought by Ruth based on the Age Discrimination in Employment Act? What additional work team-related issues might arise related to such a potential legal challenge if Ruth's attorney filed a lawsuit?

Key Terms

Employment law

Jurisdiction

Americans with
  Disabilities Act (ADA)

Unemployment
  compensation benefits

Garnish(ment)

Minimum wage

Labor union

Interstate commerce

Child labor

Title VII

Equal Employment
  Opportunity
  Commission (EEOC)

Sexual harassment

Bona fide occupational
  qualification (BFOQ)

Reasonable
  accommodation

Ordinance

Protected class

Living wage

Operator's 10-Point Tactics for Success Checklist

Evaluate your need for, and the current status of, each of the following operational tactics. For those tactics you think are important, but not yet in place, develop an action plan for its implementation including who will be responsible for the tactic's completion and the target date by which it should be completed.

Tactic	Don't Agree (Not Done)	Agree (Done)	Agree (Not Done)	If Not Done Who Is Responsible?	If Not Done Target Completion Date
1) Operator recognizes the importance of knowing and following the laws and ordinances related to the employment of workers.	___	___	___		
2) Operator can state the purpose of the Clayton Act and the Railway Labor Act (RLA).	___	___	___		
3) Operator can state the purpose of the National Labor Relations Act and the Fair Labor Standards Act (FLSA).	___	___	___		
4) Operator can state the purpose of the Equal Pay Act and the Age Discrimination in Employment Act.	___	___	___		

Tactic	Don't Agree (Not Done)	Agree (Done)	Agree (Not Done)	If Not Done	
				Who Is Responsible?	Target Completion Date
5) Operator can state the purpose of the Civil Rights Act and the Title VII features of that act.	——	——	——		
6) Operator can state the purpose of the Pregnancy Discrimination Act and the American with Disabilities Act (ADA).	——	——	——		
7) Operator can state the purpose of the Family Medical Leave Act (FMLA), and the Patient Protection, and Affordable Care Act (ACA).	——	——	——		
8) Operator recognizes the importance of knowing and following all state laws related to employee management.	——	——	——		
9) Operator recognizes the importance of knowing and following all local laws and ordinances related to employee management.	——	——	——		
10) Operator recognizes the importance of monitoring and, where appropriate, providing input into proposed employment legislation that affects their segment of the foodservice industry.	——	——	——		

3

Workplace Health and Safety

What You Will Learn

1) The Importance of Employee Health, Safety, and Security
2) How to Help Protect Employee Health
3) How to Help Ensure Employee Safety
4) How to Optimize Employee Security

Operator's Brief

The presence of qualified employees is key to the success of any foodservice operation. In this chapter, you will learn that the best foodservice operators carefully design and implement programs to help ensure the health, safety, and security of their staff members.

While it is certainly good business for you to take appropriate steps to help protect your employees at work, it is also a legal requirement that you do so. In this chapter, you will learn about important legislation related to protecting worker health as well as the many financial benefits of doing so.

Your actions as an employer will go a long way toward protecting your workers' health. A variety of such actions are addressed in this chapter including employee assistance programs and wellness programs. These are two examples of affirmative programs that can be implemented to assist in maintaining your employees' health.

Healthy workers must, of course, also be kept safe while working. The Occupational Safety and Health Administration (OSHA) is the federal agency responsible for helping to ensure employees' safety, and, in this chapter, you will learn about their efforts as well as safety planning activities you can undertake on your own to help maintain a safe workplace.

Even when a workplace is healthy and safe, there can still be internal and external threats to worker security. Harassment of all types (and its prevention in the workplace) is one of the most essential security-related areas for you to know about. Numerous ways to minimize the potential for harassment threatening workers, as well as other forms of workplace violence are detailed in this chapter.

CHAPTER OUTLINE

The Importance of Protecting Team Members
    Legal Aspects of Employee Protection
    Financial Aspects of Employee Protection
Creating a Healthy Workplace
    Employee Assistance Programs
    Employee Wellness Programs
Maintaining a Safe Workplace
    The Occupational Safety and Health Administration (OSHA)
    Safety Planning
Maintaining a Secure Workplace
    Preventing Harassment
    Preventing Workplace Violence

The Importance of Protecting Team Members

Most foodservice operators agree they have an obligation to ensure their workplaces are free from unnecessary hazards and that workplace conditions are conducive to their employees' physical and mental health. In addition to concerns about worker health, foodservice operators must also be concerned about employee **safety** and **security**.

Key Term

Safety (employee): A condition that minimizes the risk of harm to employees.

Key Term

Security (worker): Employees' feelings about fear and anxiety.

In many English-language thesauruses, readers find the words *safety* and *security* listed as synonyms (because they both have the same general meaning). However, for workers in the foodservice industry, each designates a distinct concept. For example, a worker in an operation might be very safe but still not feel secure. Alternatively, workers might feel they are secure when they are, in reality, quite unsafe but not aware of it.

Foodservice operators concerned about the well-being of their employees recognize that there are two important concepts:

1) Legal aspects of employee protection
2) Financial aspects of employee protection

Legal Aspects of Employee Protection

Despite the importance to society (and businesses) of protecting members who must work, the first legislation specifically designed to address workplace safety was not enacted in the United States until 1970. In that year, Congress passed the **Occupational Safety and Health Act** and, in doing so, it created the **Occupational Safety and Health Administration (OSHA)**.

The passage of the Occupational Safety and Health Act dramatically changed the way foodservice operators viewed their role in ensuring that the physical working conditions in their operations meet subscribed standards. The Civil Rights Act of 1964 (see Chapter 2) significantly altered how employees were to be selected and treated while at work. In much the same way, passage of the Occupational Safety and Health Act ultimately altered the physical conditions in which workers do their jobs.

Most members of society work to maintain the lifestyle they desire for themselves and their families. Therefore, it is not surprising that society maintains a legitimate interest in requiring employers to take reasonable steps to ensure worker safety. Some business owners and operators might find legislation related to employee safety to be time-consuming and/or cumbersome to implement. However, it is in the best interests of all businesses to minimize on-job accidents, and this is especially so when reasonable management actions could prevent worker accidents or deaths.

Key Term

Occupational Safety and Health Act: The law that gave the Federal Government the authority to establish and enforce safety and health standards for most employees in the United States.

Key Term

Occupational Safety and Health Administration (OSHA): The agency of the Federal Government that assures safe and healthful working conditions by establishing and enforcing standards, and by providing necessary training, outreach, education, and assistance.

Financial Aspects of Employee Protection

Providing for the health, safety, and security of an operation's team members is not just the legal thing to do; for businesses it is also the financially sensible thing to do as well. There are numerous cost-affected areas directly related to workplace health and safety. While the actual expense to be incurred depends on the specific situation, some of the most important direct, indirect, and intangible costs associated with worker health and safety issues include:

✓ *Accident-related costs:* Unnecessary accidents and illness lead directly to increased employee costs. Examples include lost productivity when experienced workers are replaced with inexperienced workers, and overtime costs paid to employees who cover the shifts of injured workers. Also, increased stress often occurs to workers who must make up for the output of their absent peers.

✓ *Recruitment and retraining costs:* When workers are severely injured or are ill over an extended time, new workers must often be recruited. These costs can often be significant when new workers must be trained, and this typically requires expense, time, and effort on the part of an operation.

✓ *Legal expense:* Workers who are injured on the job due to carelessness or negligence on the part of an employer may have a sound basis for legal action taken against the employer. *Note:* This can often be true for guests as well. Even if operators can successfully defend their actions, significant legal fees in defense of operating policies and procedures can be incurred.

✓ *Facility-related costs:* In some cases, accidents can cause an operation to undergo facility repair-related costs. Minor fires, injuries resulting from equipment broken during improper use, and damaged furniture, fixtures, and equipment (FFE) often result from employee accidents.

✓ *Negative publicity:* Some serious situations such as a highly publicized robbery, fire, workplace violence, and/or employee accidents can generate substantial negative publicity. This, in turn, can diminish an operation's reputation in the community and depress its sales.

✓ **Workers' compensation fund** *costs:* All food-service employers must pay into their state's workers' compensation insurance fund. These systems provide business owners with a significant amount of control over the cost of their workers' compensation premiums: safe businesses are rewarded with lower premiums and unsafe businesses pay higher premiums.

While the above list of costs related to worker health and safety is not intended to be exhaustive, it is clear that operators who do not diligently address issues of worker health, safety, and security can face significant costs as a result.

Key Term

Workers' compensation fund: A state-operated program that provides medical expenses, lost wages, and rehabilitation costs to employees who are injured or become ill "in the course and scope" of their job. It also pays death benefits to families of employees who are killed on the job. Payments are made into the fund by state-mandated employer premiums.

Creating a Healthy Workplace

Unhealthy workplaces should be of concern to all foodservice operators. Guests and the entire operation suffers when employees cannot function properly because of related headaches, watering eyes, nausea, or fear of exposure to elements that can cause long-term health problems. Consequently, maintaining a healthy work environment benefits all foodservice operations, their work teams, and their guests.

Although the specifics of exactly what constitutes a healthy work environment can vary based on individual foodservice operators, concerns about employee health should always ensure that the workplace:

1) *Provides enough fresh air:* In the foodservice industry, air quality concerns in work areas can be significant. Ventilation hoods in cooking and other food production areas should provide workers with enough fresh air to do their jobs comfortably.

2) *Keeps air ducts clean and dry:* Water and grease in air ducts is a fertile breeding ground for mold, mildew, and fungi. Regularly cleaning of air ducts and filters can help prevent air quality problems such as these before they start.

3) *Maintains effective equipment inspection programs:* The frequent and thorough inspection of foodservice equipment, especially items using gas as their energy source (e.g., hot water heaters, boilers, ovens, ranges, and broilers), can help detect gas or carbon monoxide leaks before they endanger workers or guests.

4) *Monitors repetitive movement injuries:* When workers must do repetitive tasks, they risk the potential for headaches, swollen feet, back pain, or nerve damage. The most frequent site of repetitive movement injury is the wrist due to *carpal tunnel syndrome.* Properly designed work areas can help minimize or eliminate repetitive stress injuries of this type.

5) *Monitors stress levels:* Stress can be caused by a variety of factors, but an in-depth discussion of stress, various means of reducing it, and the many methods of channeling stress in a positive direction is beyond the scope of this book. Operators must be aware, however, that, in the fast-paced foodservice industry, stress can be job-related, and/or it can be brought to the job.

   Workers often arrive at work concerned about personal matters such as troubling family issues, personal economic problems, and even their individual personality characteristics (e.g., prone or not prone to have high-stress levels). When the stress levels at work are excessive, absenteeism, burnout, and employee turnover rates typically increase.

6) *Provides first aid training:* OSHA requirements mandate that, in the absence of a health care facility near to the workplace that is used for the treatment of

all injured employees, a person(s) shall be adequately trained to render first aid. In addition, many operations ensure they have one or more employees trained to administer cardiopulmonary resuscitation (CPR) to fellow workers or guests. *Note:* The American Red Cross provides this type of CPR training.

7) *Pays attention to workers' complaints:* Dates, events, and employee concerns should be regularly recorded to detect any potential patterns that indicate continued threats to employee health.

Employee Assistance Programs

Even when foodservice operators work hard to provide healthy worksites, employees will occasionally have personal problems. Whether the problem relates to job stress or to marital, relationship, legal, financial, and/or health issues, the common feature of such issues is simply this: the issue will eventually be reflected at the workplace in terms of lowered productivity levels and increased absenteeism or turnover.

To help employees address problem areas in their lives, some operators implement **employee assistance programs (EAPs).**

An EAP relates to a variety of employer-initiated efforts to assist employees with family concerns, legal issues, financial matters, and health maintenance. The identifiable goals of these programs benefit operators (to return employees to productive work when possible) and employees who are bothered by the initial problem(s).

Key Term

Employee assistance programs (EAPs): The term that describes a variety of employer-initiated efforts to assist employees in the areas of family concerns, legal issues, financial matters, and health maintenance.

To illustrate the importance of EAPs, consider the case of Mark Chaplin, a cook in a popular restaurant. Mark is a talented worker and has performed well during his 15 years of employment with the operation, and he is respected by his fellow employees and supervisor.

Most recently, however, Mark's performance has not been so good. He has been late three times in the past five weeks, and, just yesterday, he called in sick to work only 15 minutes before the beginning of his scheduled shift. The rumor in the restaurant is that Mark is having significant marital problems and that his wife and children are now living in a town several hours away. The operator of the business is considering what to do.

Mark could, of course, be legally suspended or terminated if he is found to be in violation of the restaurant's attendance policy. Ultimately terminating Mark, however, will likely result in a new employee search involving significant expenses of time and money and an extended period of training to bring the new employee up to Mark's level of experience and productivity. In this case, the operator may prefer to help Mark through this difficult personal period and help to retain him as a high-quality employee. Effective EAPs are designed and implemented for cases such as Mark's.

Today some hospitality employers provide their workers help in nontraditional areas of support such as adoption counseling, legal assistance, and bereavement counseling, as well as mental health and substance abuse counseling. These programs can be very cost-effective if the employees view them positively. For employees, the biggest concern about utilizing an employer-provided EAP relates to confidentiality. It is important to note that EAP services are administered by professionals with experience in worker assistance, not the foodservice operators utilizing their services. Employees must be assured that, if they voluntarily enroll in an EAP, their participation will be kept strictly confidential. In high-quality EAPs, the professional administrators of the programs ensure that the confidentiality of its participants is scrupulously maintained.

Employee Wellness Programs

Some progressive operators take the position that it is best to help employees eliminate lifestyle factors that can lead to personal problems and, as a result, can help prevent some difficulties before they begin. To do so, these operators may implement an employee **wellness program**.

Typical examples of employer-initiated wellness programs include the topics of smoking cessation, nutrition, weight loss and management, high blood pressure control, self-defense, exercise, and stress management.

Key Term

Wellness program (employee): An employer-sponsored initiative designed to promote the good health of employees.

Foodservice operators who provide these programs often find that their employees stay healthier and that the business's health insurance providers offer discounts for implementing these types of programs. In addition, some operators find that employee participation in wellness programs increases when the employee's family members participate because the rules allow them to do so. For example, workers who are interested in participating in a company-sponsored weight loss program may be more inclined to regularly attend sessions if their spouse, partner, significant other, or family members can also join in.

Maintaining a Safe Workplace

All employees in a foodservice operation deserve a safe workplace. A safe and healthy workplace protects workers from injury and illness. It can also lower a foodservice operator's injury and illness-related costs, reduce absenteeism and employee turnover rates (see Chapter 1), increase productivity and quality, and raise employee morale.

Worker safety is good for business and protecting workers is the right thing to do. To best help protect the safety of team members, foodservice operators must fully understand the role of OSHA in ensuring worker safety and their own responsibilities for safety planning.

The Occupational Safety and Health Administration (OSHA)

When the U.S. Congress passed the Occupational Safety and Health Act, it established for the first time, a nationwide and federal program to address worker safety. It is important to note that foodservice operators who employ 10 or fewer workers may not be required to follow all OSHA regulations. However, it is also good to recognize that most OSHA requirements are designed to keep workers safe. As a result, even if an operator is exempt from some of the regulations, it is most often still a good idea to follow them because compliance can help the operator's workers stay safer at work.

As initially created, OSHA's stated mission was to prevent work-related injuries, illnesses, and deaths. OSHA's current role is to ensure the safety and health of America's workers by setting and enforcing standards; providing training, outreach, and education, establishing partnerships, and encouraging continual improvement in workplace safety and health.

The work of OSHA has been a tremendous success. For example, worker deaths in America are down, on average, from about 38 worker deaths a day in 1970 to 13 a day in 2020. Worker injuries and illness are also down from 10.9 incidents per 100 workers in 1972 to 2.7 per 100 incidents in 2020.

The safety and health standards administered by OSHA are complex. They include standards related to noise levels, air quality, physical protection equipment, and proper size for ladders, to name but a few. Regardless of one's view of the detail with which OSHA involves itself, foodservice operators are responsible for knowing and following the act's provisions. From a practical perspective, foodservice operators most often interact with OSHA in the areas of compliance and recordkeeping.

OSHA requires employers to keep detailed records regarding employee illnesses and accidents related to work and to calculate on-job accident rates. OSHA monitors workplace safety with a large staff of inspectors called compliance officers. These officers visit workplaces during regular business hours and can perform unannounced inspections to ensure employers are operating in compliance with OSHA health and safety regulations.

In addition, the officers are required to investigate any complaints of unsafe business practices. Foodservice operators can accompany OSHA compliance officers during an inspection of their facility, and they do so for two reasons: (1) the operator may be able to answer questions or clarify procedures for the compliance officer and (2) the operator should know what transpires during the inspection. Afterward, an operator should discuss inspection results with the compliance officer and request a copy of inspection reports filed.

In the foodservice industry, some of the most commonly cited safety violations and penalties relate to an employee's "right to know" about potential threats to their safety. Early in 1984, OSHA put in place the Federal Hazard Communication Standard now known as the "right-to-know" law.

Originally, the law affected primarily chemical manufacturing facilities. In 1985, however, the courts decided that these regulations should apply to all facilities. The right-to-know law is designed to protect workers from potentially hazardous chemicals. The requirements and regulations concerning the right to know include the mandatory use of **safety data sheets (SDSs)** previously called material safety data sheets (MSDS).

Key Term

Safety data sheets (SDSs): Documents that describe how to safely use, store, and dispose of potentially hazardous chemicals.

A SDS must be distributed to hazardous chemical buyers at the time of the first shipment of a covered product. Employers must share the information contained in SDSs with the employees who will use the hazardous chemicals. Employers must also ensure that SDS materials are written in English and are readily accessible to the employees in their work areas throughout each work shift.

Many foodservice operators employ workers who speak no English or whose English skills are very limited. OSHA rules in these cases require that an employer must instruct workers using both a language and vocabulary that the worker can understand. For example, if a worker does not speak or comprehend English, instruction must be provided in a language the employee can understand.

Similarly, if the employee's vocabulary is limited, proper safety training must account for that limitation. By the same token, if employees are not literate, telling them to read training materials will not satisfy an employer's safety training obligation.

Foodservice operators must generally realize that, if they customarily communicate work instructions or other workplace information to employees at a certain vocabulary level or in language other than English, they will also need to provide safety and health training to employees in the same manner.

The Occupational Safety and Health Act also requires that the Secretary of Labor produce regulations that ensure employers keep records of occupational deaths, injuries, and illnesses. Recording or reporting a work-related injury, illness, or fatality does not mean that the employer or employee was at fault, that an OSHA rule has been violated, or that the employee is eligible for workers' compensation or other benefits.

OSHA uses injury and illness statistics to help direct its programs and measure its own performance. Inspectors use the data during inspections to help direct their efforts at hazards that hurt workers. The records employers submit to OSHA also provide the base data for the Annual Survey of Occupational Injuries and Illnesses which is the country's primary source of occupational injury and illness data.

Foodservice operators who must follow OSHA rules are required to keep records of their safety-related efforts and results. Currently, the major areas of OSHA-mandated record keeping standards related to the foodservice industry include:

✓ *Log and summary of all recordable injuries and illnesses*—All OSHA-recordable injuries and illnesses that occur in the workplace or as part of the employee's duties must be entered in a log approved by OSHA within six working days after the employer is notified that a recordable injury or illness has occurred. A summary of the reported accidents or illnesses must be signed by a responsible company official and posted in regular work areas.

✓ *Personal protective equipment* (*assessment and training*)—The Personal Protective Equipment (PPE) standard addresses an employer's responsibility to identify potential threats to an employee's eyes, face, head, and extremities. It also allows for the necessary clothing or gear required to protect employees from harm. If potential hazards including chemical or radiological hazards or mechanical irritants are identified, issues related to those substances must be assessed. The assessment must be written, certified by a responsible official, and be work area and job specific.

Key Term

Personal protective equipment (PPE): Equipment worn to minimize exposure to hazards that can cause serious workplace injuries and illnesses.

✓ *Control of hazardous energy (lock out/tag out)*—The Lock Out/Tag Out standard applies to activities related to servicing and/or maintaining machines and equipment. It is intended to protect employees from the unexpected movement or start-up of machines and equipment. OSHA requires that employers develop a written plan that identifies specific equipment and activities that would require equipment lock-out or tag-out when broken or under repair and include an employee training element about them.

✓ *Hazard communication standards*—The Hazard Communication standard addresses the issue of potentially hazardous chemicals in the workplace and informing employees about specific hazards and protective measures to be used when operating, handling, and/or storing these products. Some recordkeeping requirements of this standard include (1) developing and maintaining a written hazard communication program including lists of hazardous chemicals in the workplace; (2) providing employees with access to SDSs; (3) employee training (including documentation) about the hazards of the chemicals they are or may be exposed to and for which protective measures must be taken; and (4) labeling of chemical containers in the workplace.

✓ *Emergency action plans and fire prevention plans*—Any facility that employs more than 10 people must develop a written Emergency Action and Fire Prevention Plan. Facilities that employ fewer than 10 people do not have to develop a written plan, but they are required to orally communicate emergency action procedures to each employee.

Additional examples of recordkeeping requirements that may or may not directly apply to a specific foodservice operation include those related to the respiratory protection of workers, asbestos exposure, hepatitis B, and blood-borne pathogens. Foodservice operators can remain current about recordkeeping requirements by regularly logging into the OSHA website (www.osha.gov).

Technology at Work

Accurate workplace injury and illness recordkeeping and reporting is crucial for employers, workers, and OSHA in evaluating workplace safety, understanding potential hazards, and implementing training and/or protections to ensure safe and healthful working conditions.

Many foodservice operators recognize the importance of maintaining accurate OSHA-required records, and they select specific software to help them do so.

Fortunately, several companies offer inexpensive software that allows foodservice operators to record important incidents and report required data to OSHA on a mandated schedule.

To review some of these companies' product offerings, enter "OSHA reporting requirement software for restaurants" in your favorite search engine and view the results.

What Would You Do? 3.1

Carlos Magana was a non-English-speaking dishwasher/janitor working in a health care facility's kitchen. Bert Lagrange was the new Food & Beverage Director at the facility.

Bert, who spoke only English, instructed Mr. Magana to clean the grout between the red quarry kitchen tiles with a powerful cleaner that Mr. Lagrange had purchased from a chemical cleaning supply vendor.

Bert demonstrated to Mr. Magana (who spoke only Spanish) how Carlos should pour the chemical directly from the bottle onto the grout, and then brush the grout with a wire brush until it was white.

Since the cleaner was so strong and because Carlos was not wearing protective gloves, his hands were seriously irritated by the chemicals in the cleaner. To lessen the irritation to his hands, Carlos decided, on his own, to dilute the chemical.

He added water to the bottle of cleaner, not realizing that the addition of water would cause toxic fumes. Carlos inhaled the fumes while he continued cleaning and later suffered serious lung damage.

Bert was subsequently contacted by OSHA, which cited and heavily fined the facility for a safety data sheet violation despite the fact that Bert's employer maintained that SDS documents, including the one for the cleaner in question, were, in fact, available for inspection by employees.

Assume you were Bert and that the proper SDS was, in fact, available (in English only) to the facility's employees. Do you believe you had a legal obligation to provide safety information to Mr. Magana in his primary language (Spanish)? A moral obligation? How should you have done it? Explain your answer.

Safety Planning

Foodservice operators should ensure their employees are as safe as possible at work, and, as a result, they feel a high degree of security. Operators can do so by implementing well-designed programs to enhance worker safety and security. For example, an operator could implement fire safety programs designed to minimize the risk of fire-related injuries to employees.

Figure 3.1 lists additional types of threats and **crises** for which an operator might consider designing specialized employee safety and/or security programs.

Since the safety and security needs of different foodservice operations vary widely, providing one all-purpose, step-by-step list of implementa-

Key Term

Crises (work): Situations that have the potential to negatively affect the health, safety, or security of employees.

tion activities that minimizes opportunities for employee accidents or injury is difficult. From a legal perspective, however, an operator's basic obligation is to act responsibly in the face of threats. Operators can analyze and respond to these responsibilities using a four-step method:

Step 1: Recognize the safety threat. Safety programs generally start with recognizing a need such as realizing a threat to people (or property) exist. Some of the

Vandalism	Looting	Ice	Accidental death
Fire	Hurricanes	Accident/injury	Suicide
Arson	Tornadoes	Drug overdose	Civil disturbance
Bomb threat	Floods	Medical emergency	Terrorist attack
Robbery	Snow storms	Breathing emergency	Foodborne illness

Figure 3.1 Potential Threats to Worker Safety

most common threats to employee safety in the foodservice industry include those related to natural disasters, co-workers, guests (in dining rooms, bars, lounges, and the like), the worksite, and even the employees themselves.

Step 2: Develop a program in response to the threat. Once a threat to safety or security is identified, operators can develop an appropriate response to address it. The proper response to an identified threat may take the form of any one or more of the following:

1) Employee training for threat prevention
2) Increasing surveillance and/or patrol of facilities (for external threats)
3) Systematic inspections
4) Modifying physical facilities to reduce the threat
5) Establishing standard operating procedures (SOPs) (see Chapter 1) to address the threat(s)

Step 3: Implement the program. When a foodservice operator has identified a threat and designed a safety and security program to directly address it, they must put the program into action. For many operators, a property safety committee can play a valuable role in identifying and correcting safety and security problem areas. Ideally, a safety committee will consist of members from each of the operation's departments.

For example, a foodservice operation might have members from the preproduction, production, and clean-up areas in the back of the house and bartenders, servers, and hosts from the front of the house. In very small foodservice facilities, the operator of the business assumes the primary responsibility for ensuring the implementation of appropriate safety-related training.

Step 4: Evaluate the program. If a safety program is not working (it is not reducing or eliminating an identified threat), then the program must be reviewed for modification. If legally necessary, an operation is in a stronger position to defend safety-related activities if it can document not only a safe program, but also an effective program.

There are several ways to measure a safety program's effectiveness and some tangible measurements operators might use include:

✓ Number of facility inspections performed
✓ Number of safety-related incidents reported
✓ Dollar amount of losses sustained
✓ Number of insurance claims filed
✓ Number of lawsuits filed
✓ Number of serious or minor accidents
✓ Number of workdays lost by employees
✓ Number of drills or training exercises properly performed

The important point to remember is that a safety-related program can only be successfully implemented when an appropriate evaluation component for the program has been developed and implemented.

Find Out More

One of the best ways to reduce costs, retain jobs, and maintain a productive workforce is to reduce injuries. One of the best strategies to prevent job injuries is regular health and safety training for foodservice employees.

Regular safety training helps workers learn how to avoid hazards, keeps lines of communication open between team leaders and team members and lets workers know that management is serious about promoting sound safety policies and work practices.

The training needs of a specific foodservice operation vary based upon the products it produces and the manner in which these products are served. Most foodservice operations, however, have some common safety training needs. These include:

Preventing burns
Preventing cuts
Preventing injuries from slips and falls
Preventing injuries from ergonomic (repetitive motion) hazards
Planning for fire emergencies on the job
Dealing with injuries on the job
Assisting guests in medical emergencies

Many entities including OSHA, state employment agencies, professional trade associations, and for-profit training organizations provide free or inexpensive foodservice safety training materials. To review some of these valuable instruction and training tools enter "Safety training for restaurant workers" in your favorite search engine and review the results.

Technology at Work

In the not too recent past, the use of video cameras to capture activity in security sensitive areas such as foodservice operations' parking lots, back door entrances, and administrative offices, was cumbersome. Physical videotapes needed to be changed, dated, and stored, and often these procedures were not properly performed.

Video security is still important. For example, studies have shown that restaurants without proper security systems are more likely to be robbed than those with a well-thought-out security system and burglar alarms. Proper video surveillance not only protects a business, but its staff and customers as well.

Today, there have been tremendous advancements in ease and reduced costs for providing appropriate video surveillance systems. Most systems can send a text to alert a foodservice operator about unusual activity. They can also include smart tracking systems that enable a camera to keep track of moving objects detected by motion sensors and immediately upload results to the cloud.

To review some of many advancements in video security systems enter "video security systems for restaurants" in your favorite search engine and view the results.

Maintaining a Secure Workplace

It would be nice to believe that groups of employees, all united toward achieving the same goals, work with a high degree of peace, harmony, and security. Unfortunately, sometimes that simply is not the case. Although hospitality workplaces are generally peaceful, foodservice operators must take steps to ensure, to the greatest degree possible, that they stay that way.

As previously addressed, foodservice operators must take a variety of steps to help ensure their employees enjoy a high degree of health and safety while at work. All team members deserve an operator's best efforts to ensure their security as well. In most cases, these security-related efforts take the form of protecting workers from harassment and workplace violence.

Preventing Harassment

If foodservice workers are to enjoy security (freedom from fear and anxiety) at work, they must work in a harassment-free environment. While in the past much media attention has focused on sexual harassment (see Chapter 2), eliminating all forms of harassment (e.g., racial, gender, sexual orientation, religion, and others) from the workplace is one of a foodservice operator's most important jobs.

Today there is greater understanding among foodservice operators that the Civil Rights Act prohibits sexual (and other types of) harassment at work, even when the harassment occurs among people of the same gender. In addition, most states have their own laws regarding fair employment practices that prohibit harassment based on a variety of factors; as a result, many state harassment laws are stricter than the federal law.

Employees can face a variety of harassment forms including:

✓ *Bullying*—Harassment that can occur on the playground, in the workforce, or any other place. Usually, physical and psychological harassing behavior is perpetrated against an individual by one or more persons.

✓ *Psychological harassment*—Humiliating or abusive behavior that lowers a person's self-esteem or causes them torment. This can take the form of verbal comments, actions, and/or gestures.

✓ *Racial harassment*—The targeting of an individual because of their race, origin, or ethnicity. The harassment can include words, deeds, and actions that are specifically designed to make the target feel degraded because of their race, origin, or ethnicity.

✓ *Religious harassment*—Verbal, psychological, or physical harassment used against targets because they choose to practice a specific religion.

✓ *Stalking*—The unauthorized following and surveillance of an individual to the extent that the person's privacy is unacceptably intruded upon, and the victim fears for their safety. In the workplace, those who know, but do not necessarily work with, the victim most commonly exhibit this form of harassment.

✓ *Sexual harassment*—This harassment can happen anywhere but is most common in the workplace. It involves unwelcome words, deeds, actions, gestures, symbols, or behaviors of a sexual nature that make the target feel uncomfortable. Gender and sexual orientation issues are this form of harassment.

Harassment is any unwelcome conduct on the job that creates an intimidating, hostile, or offensive working environment. Such environments are unsafe for employees both physically and mentally. In real-life terms, harassing behavior ranges from repeated offensive or belittling jokes to outright physical assault. In the hospitality industry, workers can be subject to extreme forms of harassment by co-workers and customers.

Workers can be required to follow company policies regarding harassment, but employers can also be held directly responsible for the acts of customers and vendors who are not subject to the company's disciplinary procedures. Therefore, the effective management of a **zero tolerance** policy should be implemented, should apply to

Key Term

Zero-tolerance (policy): A policy that permits no amount of leniency regarding harassing behavior.

all individuals with whom employees come into contact, and should be designed to help ensure the safety and security of all employees.

Since there is more established case law about how the courts view sexual harassment, and because of its common occurrence in the hospitality industry, the major focus of the following section relates to preventing sexual harassment. Many of the prevention principles addressed, however, also apply to other forms of harassment.

The U.S. Supreme Court first ruled that unwelcome sexual conduct creating a hostile and offensive work environment violates Title VII of the Civil Rights Act in *Meritor Savings Bank, FSB v. Vinson* (1986). Since then, the EEOC (see Chapter 2) and the courts have expanded the definition of a hostile and offensive work environment to prohibit harassment based on race, color, sex (including sexual orientation), religion, national origin, age, and disability.

The regulations issued by the EEOC on sexual harassment provide that

> *[a]n employer should take all steps necessary to prevent sexual harassment from occurring, such as affirmatively raising the subject, expressing strong disapproval, developing appropriate sanctions, informing employees of their right to raise and how to raise the issue of harassment under Title VII, and developing methods to sensitize all concerned.*[1]

The EEOC expects employers to be affirmative (not just reactive!) when taking steps to prevent all types of harassment. It is also important to understand that the *Meritor Savings Bank* case ruling by the Supreme Court addressed two different standards for determining liability in cases of **quid pro quo (sexual harassment)** and **hostile work environment**.

Key Term

Quid pro quo (sexual harassment): Quid pro quo literally means "something for something." This harassment occurs when a supervisor behaves in a way or demands actions from an employee that force the employee to decide between giving in to sexual demands or losing their job, losing job benefits or promotion, or otherwise suffering negative consequences.

Key Term

Hostile work environment: A workplace infused with intimidation, ridicule, and insult that is severe or pervasive enough to create a seriously uncomfortable or abusive working environment with conduct severe enough to create a work environment that a reasonable person would find intimidating (hostile).

---

1 https://www.ecfr.gov/current/title-29/subtitle-B/chapter-XIV/part-1604/section-1604.11. Retrieved May 5, 2023.

From a legal perspective, if harassment is established under the quid pro quo version, an employer is *automatically* liable and will be held accountable whether or not steps were taken to correct the situation. By contrast, an employer's liability in a hostile work environment case must be established by showing not only that the harassment occurred but also that the employer did not take appropriate action to stop it.

Sexual (and other forms of) harassment policies and the training procedures required to fully comply with federal and state laws can be complex and should be thoroughly reviewed by legal counsel before their implementation. For general guidelines in preventing harassment of all types, however, operators must understand:

✓ *What is, and is not, a hostile work environment*—A hostile work environment can exist even if no employees have yet complained. As a result, operators should stop any actions that could generate charges of harassment.

✓ *The company policy*—An operation's harassment policy should be familiar to all workers and their supervisors. Sometimes, supervisors may know there is a harassment policy but have little idea of what it entails. If the operation is sued and supervisors cannot display a sufficient understanding of the company policy, then it will likely be viewed by the courts as not having been enforced.

✓ *The effect of speech*—Nearly everyone working in foodservice has heard coworkers or customers use language that could be considered inappropriate, hostile, sexist, racist, ethnically charged, crude, gross, insensitive, age-based, or derogatory. Foodservice operators need to identify and inform all team members and their supervisors about the specific types of language that will likely lead to charges of harassment in their own businesses.

✓ *Proper investigations*—Properly conducted harassment investigations are essential to limiting an operation's legal liability. An operation's harassment policy should make clear who investigates claims, what should be done if a supervisor is the target of the investigation, at what point legal counsel should be involved, and whether (when applicable), a specific supervisor can continue to oversee the complaining employee during the investigation.

✓ *Personal liability*—When a harassment case results in a lawsuit, the operation is sued, but in many cases so is the individual responsible for managing it. This is possible even though that individual was not the harasser, did not know about the situation, and the employee allegedly being harassed did not report it.

The concept of zero-tolerance (harassment) policies should address when the language used, or action displayed constitute harassment. This question can be difficult to answer because not all employees would find the same behavior offensive. For example, some culinary employees (male and female) might consider hearing coworkers swearing while working the entrée station in the kitchen

during a busy Saturday night rush as very offensive and therefore potentially harassing, when
others may not find it to be so.

Key Term

**Reasonable person
standard:** The typical or
average person (and their
behavior and beliefs) placed
in a specific environmental
setting.

The courts have addressed this question and
have instituted a **reasonable person standard**.
When determining whether conduct is considered
hostile environment harassment, the courts, as
well as the EEOC, typically evaluate the objectionable conduct from the standpoint of a "reasonable
person" under similar circumstances. Under this standard, complaining employees
must establish not only that they personally perceived their work environment to be
hostile but also that a reasonable person would have perceived it to be hostile as well.

Foodservice operators must also understand the implications of the court's
increasing acceptance of a "reasonable woman" standard to use instead of the
reasonable person standard for sexual harassment cases. In *Ellison v. Brady* (1991),
the U.S. Ninth Circuit Court concluded that offensive conduct must be evaluated
from the perspective of a reasonable person of the same gender as the victim. That
court rejected the non-gender-specific reasonable person standard because it does
not take into consideration the different perspectives of men and women. In its
ruling the court wrote,

> *A complete understanding of the victim's view requires, among other things,
> an analysis of the different perspectives of men and women. Conduct that
> many men consider unobjectionable may offend many women.*[2]

Everyone should agree that harassment of any type has no place in a work environment. Operators must also understand, however, that a *charge* of sexual harassment is not synonymous with an incident of sexual harassment any more than
being accused of a crime means the accused is automatically guilty of it. With
perceived abuses by some who have charged harassment and the response of
some employers, the state and federal courts are increasingly defining the rights
of the accused in sexual harassment cases. However, these rights vary by state. In
Ohio, for just one example, state law requires that the "accused" have the right:

1) *To be free from discrimination*—For example, a member of a minority group
   may not be treated differently from whites when assessing whether harassment occurred or how it will be dealt with.
2) *To a thorough investigation*—An employer cannot conduct a "kangaroo court"
   (i.e., a mock or unauthorized justice proceeding) without risking a jury

---

2 https://casetext.com/case/ellison-v-brady. Retrieved May 5, 2023.

second-guessing what the employer might have found if it had looked at all the facts. It is simply naïve to believe that there have never been unfounded charges of harassment.

3) *To a good-faith basis for believing that the allegations are true before taking adverse employment action*—This is especially true if the employee can point to others who were treated differently for the same alleged offense.

4) *To be free from defamation*—An employer should not share information about an investigation with anyone other than those who need to know the results.

Employers have a clearer duty to protect employees from harassment, which carries far greater penalties if it is breached, than to protect the rights of the accused. Therefore, in harassment cases that are not clear-cut, operators are best advised to continue to make errors on the side of protecting the victim. Increasingly, however, employers are legally required to keep the rights of the accused in mind, especially as the number of same-sex harassment cases wind their way through the court systems.

Key Term

Workplace violence: Any act in which a person is abused, threatened, intimidated, or assaulted in their place of employment.

Preventing Workplace Violence

While harassment is most often considered verbal abuse at work, foodservice operators must increasingly concern themselves with **workplace violence**, a concept that includes harassment, physical assault, and even homicide.

Just as some operators think of harassment only as a verbal attack, some think of violence only in terms of a physical assault. However, workplace violence is a much broader problem. It occurs any time workers are abused, threatened, intimidated, or assaulted in their place of employment. Workplace violence includes:

✓ *Threatening behavior*—Includes actions such as shaking fists, destroying property, or throwing objects.

✓ *Verbal threats*—Include **implicit threats** and **explicit threats**, made in person, or left on voice mail, posted on social media, or sent as a verbal text.

Key Term

Implicit threat: A threatening act that is implied rather than expressly stated. Example: The statement "I'd watch my back if I were you!" said in a menacing voice by one employee to another.

Key Term

Explicit threat: A threatening act that is fully and clearly expressed or demonstrated, leaving nothing implied. Example: The statement "If I see you in my work area again, I'll personally throw you out of it!" said in a menacing voice by one employee to another.

✓ *Written threats*—Consist of everything from Post-It notes to e-mails, texts, tweets, and even long letters.

✓ *Harassing activities*—Include a wide range of behaviors that demean, embarrass, humiliate, annoy, alarm, or verbally abuse a person. These behaviors are known as or would reasonably be expected to be unwelcome (including sexual harassment) and include words, gestures, intimidation, bullying, or other inappropriate activities.

✓ *Verbal abuse*—Includes swearing, insults, or condescending language.

✓ *Physical attacks*—Include hitting, shoving, pushing or kicking, rape, or homicide.

Still more examples of workplace violence include spreading rumors, playing pranks, inflicting property damage, vandalism, sabotage, armed robbery, theft, physical assaults, and psychological trauma. Additional examples include anger-related incidents such as rape, arson, and murder. It is important that foodservice operators understand that workplace violence is not limited to incidents that occur within a traditional workplace.

Work-related violence can occur at off-site business-related functions such as conferences and trade shows, at social events related to work, and in clients' offices or away from work, but resulting from work (e.g., a threatening telephone call from an employee to another employee's cell phone).

Workplace violence threatens both male and female workers. Acts of violence and other injuries are currently the third-leading cause of fatal occupational injuries in the United States. According to the Bureau of Labor Statistics Census of Fatal Occupational Injuries (CFOI), of the 5,333 fatal workplace injuries that occurred in the United States in 2019, 761 were cases of intentional injury by another person. Women made up 8.6% of all workplace fatalities, but they represented 14.5% of intentional injuries by a person in 2021.[3]

Every foodservice operation must be protected from disgruntled workers and customers. As the events of 9/11 and the increasingly regular occurring mass shootings that often target hospitality operations, workplaces must now be prepared to face traditional internal workplace threats and external threats.

Effective operators should take reasonable steps to protect their workers from violence. In the foodservice industry, these steps can include concrete activities such as:

✓ Increase workplace security by installing video surveillance, alarm systems, and door detectors.

✓ Increase lighting in dimly lit areas such as parking lots and around trash dumpsters.

---

3 https://www.osha.gov/workplace-violence#:~:text=According%20to%20the%20Bureau%20 of,%5BMore...%5D. Retrieved April 25, 2023.

✓ Trim bushes and shrubs that provide hiding places for would-be thieves and attackers.

✓ Implement effective alcohol server training programs to prevent and discourage excessive alcohol consumption (bars and restaurants).

✓ Minimize the amount of cash available to cashiers and those making bank deposits.

Specific actions taken to help deter workplace violence are important, but it is also critical to understand that, in most cases, workplace violence is committed by a business's current and former employees. Foodservice operators should write and implement workplace violence policies that apply to all current team members.

At the very least, an effective workplace violence policy addresses the following questions:

✓ What specific behaviors (e.g., swearing, intimidation, bullying, harassment, and others) do management consider inappropriate and unacceptable in the workplace?

✓ What should employees do when incidents covered by the policy occur?

✓ Who should be contacted when reporting workplace violence incidents (including a venue for reporting violent activity by one's immediate supervisor)?

✓ What should be done when threats or assaults that require immediate attention are seen (these events should be reported directly to the police at 911)?

While these points cover the basic minimum of a workplace violence policy, the best workplace violence prevention policies:

✓ Apply to ownership, management, employees, customers, clients, independent contractors, and anyone who has a "stakeholders' relationship" with the operation.

✓ Define exactly what is meant by "workplace violence" in precise and concrete language.

✓ Precisely state the consequences of making threats or committing violent acts.

✓ Encourage the reporting of all incidents of violence.

✓ Ensure that no reprisals are made against those employees reporting workplace violence.

✓ Outline the procedures for investigating and resolving complaints.

✓ Commit to providing support services to victims of workplace violence.

Find Out More

Workplace violence, whether defined narrowly to include only violent criminal acts, or more broadly to include intimidation and verbal threats, has long affected retail and service workers. Existing data, while limited, suggests that late-night retail establishments including foodservice operations, convenience stores, liquor stores, and gasoline stations, experience relatively high workplace violence rates.

By recognizing the hazards leading to violent incidents and implementing appropriate prevention and control measures, employers can improve employee safety rates. OSHA provides foodservice operators with detailed information about how to minimize the chances of workplace violence when their businesses are open late at night or all night.

To learn more about specific OSHA suggestions helpful to those operators whose employees work late-night and early-morning shifts, go to OSHA.gov. enter "Recommendations for workplace violence prevention programs in late-night retail establishments," and review the results posted on the OSHA website.

The leaders of a foodservice operation are unique individuals charged with a variety of important tasks. Foodservice operators must monitor the legal environment of employee management and provide a safe and healthy work environment. As illustrated in this chapter some of their most important tasks are deciding what their operations will do (what policies will be put in place?), and how they will do it (what procedures will they use?).

The proper development of a foodservice operation's operating policies and procedures is so critical it will be the sole topic of the next chapter.

What Would You Do? 3.2

"If he touches me again, I'm going to deck him," said Angela Larson, a cocktail server at the Windmere Casino restaurant and bar, as she walked into Peggy Kline's office. "He's just creepy," said Angela

"He" was Roger Sheets, an older man and a regular guest at the casino and its restaurant.

Peggy was the restaurant and bar manager.

Angela was an attractive, single parent of two who had worked at the restaurant for three years. Her co-workers considered her to be skilled and very friendly, and her attendance was excellent. She had served Roger many times before.

During Roger's most recent visit to the restaurant he had put his hand on Angela's back while he gave his order to her. According to Angela, his hand had trailed a good bit lower than her back before he finished placing his order.

"A couple of days after it happened, I talked to a lawyer friend who I'm dating now," Angela continued, "and he said my recent experience was sexual harassment, and I should report it now. My lawyer friend said the company needs to ban him, or I should sue, and I would win. A lot of money! So, I'm officially reporting him."

Assume you were Peggy. Based only on Angela's comments does this sound like a case of sexual harassment to you? Could it be a case of workplace violence? What do you think would be the appropriate thing for you to do in response to Angela's report?

Key Terms

Safety (employee)	Employee assistance programs (EAPs)	Quid pro quo (sexual harassment)
Security (worker)	Wellness program (employee)	Hostile work environment
Occupational Safety and Health Act	Safety data sheets (SDSs)	Reasonable person standard
Occupational Safety and Health Administration (OSHA)	Personal protective equipment (PPE)	Workplace violence
Workers' compensation fund	Crises (work)	Implicit threat
	Zero tolerance (policy)	Explicit threat

Operator's 10-Point Tactics for Success Checklist

Evaluate your need for, and the current status of, each of the following operational tactics. For those tactics you think are important, but not yet in place, develop an action plan for its implementation including who will be responsible for the tactic's completion and the target date by which it should be completed.

				If Not Done	
Tactic	Don't Agree (Not Done)	Agree (Done)	Agree (Not Done)	Who Is Responsible?	Target Completion Date
1) Operator recognizes the legal requirements related to the establishment of a healthy, safe, and secure workplace.	_____	_____	_____		

(Continued)

Tactic	Don't Agree (Not Done)	Agree (Done)	Agree (Not Done)	If Not Done	
				Who Is Responsible?	Target Completion Date
2) Operator recognizes the financial costs and benefits related to the establishment of a healthy, safe, and secure workplace.	_____	_____	_____		
3) Operator can identify specific actions that can be taken to help maintain a healthy foodservice workplace.	_____	_____	_____		
4) Operator understands the purpose and value of employee assistance programs (EAPs).	_____	_____	_____		
5) Operator understands the purpose and value of employee wellness programs.	_____	_____	_____		
6) Operator recognizes the role of the Occupational Safety and Health Administration (OSHA) in maintaining safe workplaces.	_____	_____	_____		
7) Operator can state the safety planning steps to be taken to address potential threats to safety in their own organization.	_____	_____	_____		
8) Operator recognizes the importance of protecting workers from workplace violence.	_____	_____	_____		
9) Operator recognizes the importance of preventing harassment of all types in the workplace.	_____	_____	_____		
10) Operator can identify specific steps that can be taken to help reduce internal and external threats of workplace violence.	_____	_____	_____		

4

Developing Policies and Procedures

What You Will Learn

1) The Difference between Policies and Procedures
2) The Steps Used to Develop Policies and Procedures
3) How to Create an Employee Handbook
4) The Importance of Policy and Procedure Review and Recordkeeping

Operator's Brief

To achieve their financial goals, foodservice operators rely on the identification of specific policies (goals they want to achieve) and procedures (how they will achieve the goals). In this chapter, you will learn how you can best develop policies and procedures to help you meet your own operating goals.

There are key areas of policy and procedure development that will relate directly to your workers. These include staffing your organization and developing, motivating, and maintaining your staff.

There are six key steps in the development of effective policies and procedures. These steps are:

1) Identifying the issues to be addressed
2) Considering on-site factors that affect implementation
3) Considering off-site factors that affect implementation
4) Drafting policies and procedures and submitting them for legal review
5) Developing related documentation and meeting recordkeeping requirements
6) Communicating finalized policies and procedures to affected parties

In this chapter, you will learn about these important steps.

(Continued)

When you have developed policies and procedures that you believe will assist you in attaining your goals, they should be submitted to a competent lawyer or other legal authority for a thorough review prior to their implementation. This must be done to ensure that you will be in compliance with all rules and regulations affecting your business.

After policies have been approved for use, they must be communicated to your employees. The tool used to do this is an employee handbook, and, in this chapter, you will learn how a proper employee handbook is developed.

Finally, in many cases, documents and records related to your policies and procedures must be maintained in your and/or employees' personal files for specific periods of time. In this chapter, you will learn about these requirements and how you can best ensure that you are always in compliance with applicable federal, state, and local recordkeeping regulations affecting your operation.

CHAPTER OUTLINE

Creating Policies and Procedures
    Policy and Procedure Development
    Key Areas of Policy and Procedure Development
Steps in Employee Policy and Procedure Development
Reviewing for Legal Compliance
Employee Handbooks
Recordkeeping Requirements

Creating Policies and Procedures

Foodservice operators managing their businesses must make many employee-related decisions. Just as operators know, they must consistently follow their standardized recipes if they are to consistently produce and serve high-quality menu items, they also recognize that consistency in their own employee-related standard operating procedures (SOPs; see Chapter 1) is critical to the smooth functioning of their facilities. As a result, operators can spend significant time deciding what worker-related goals they want to achieve in their operations and how they will achieve them.

At the same time, an operation's employees want to know that any rules an operator makes about employees will be equally applied to each of them. If they do not believe this is true, charges of bias, favoritism, sexism, and even racism can result. Therefore, experienced foodservice operators know they must (1) carefully design and implement their employee-related policies and (2) maintain evidence that their policies are applied fairly and consistently.

Policy and Procedure Development

To illustrate how the policy and procedures development process actually occurs, consider the case of Phyllis Nicholson. Phyllis is the owner/operator of the soon-to-be-opened Dancing Llama brew pub. Phyllis will offer a variety of locally crafted beers and a menu of sandwiches and other bar foods she believes guests will enjoy with their beverages. To open the pub, Phyllis believes she will need around 30 full and part-time employees.

To create a productive team of staff members, she will be required to make decisions about many employee-related issues and policies. A few examples include:

1) Employee selection
2) Employee scheduling
3) Required dress and uniform codes
4) Attendance and tardiness policies
5) Performance evaluation
6) Termination procedures

The actual procedures used to address these issues will be important. In this example, the procedures must address a variety of issues related to the policies Phyliss will develop, including:

1) What factors will be used to select employees?
2) How will employees be scheduled?
3) What will be the penalties for dress code violations?
4) Who will record employee absence and tardiness information?
5) How frequently will employee performance reviews be conducted?
6) What written documentation will be required in cases of employee termination?

As this example shows, there is an important relationship between what Phyllis will do (**policies**) and exactly how she will do it (**procedures**), as illustrated in Figure 4.1. These procedures may spell out rewards for policy compliance, penalties for noncompliance, and steps required for policy implementation. In many cases, a single policy will require multiple supporting procedures.

The line between what a foodservice operation will do and how it can be done can be a fine one. However, it is important to remember that to be effective an operation's policies must be supported by procedures that, when followed, ensure the fair and consistent application of the policy.

Key Term

Policy: A detailed course of action developed to guide future decision making to achieve stated objectives.

Key Term

Procedure: A technique or method used to implement a policy.

Policies: What We Will Do	Procedures: How We Will Do It
1) Select employees	1) Use standardized application forms.
2) Schedule employees	2) Make and distribute employee work schedules one week in advance.
3) Implement employee dress codes	3) Impose fair and consistent penalties for policy violations.
4) Monitor attendance and tardiness	4) Record employee arrival and departure times daily.
5) Conduct performance evaluations	5) Schedule annual employee reviews and designate the proper reviewers.
6) Terminate employees	6) Develop standards for documenting (in writing) the reason for an employee's termination.

Figure 4.1 Dancing Llama Worker Policy and Procedure Relationship

Key Areas of Policy and Procedure Development

Not all hospitality operation's policies and procedures are related directly to employees. For example, how often bank deposits should be made, how frequently should menu prices be updated, and what is the best timing to send invoice payments to vendors? All the above (and numerous others!) are examples of areas where policies and procedures are important, but they do not directly related to employees.

Employee-related policy and procedure development can, however, have a direct effect on all areas of the operation. Experienced operators know that identifying all topics within a hospitality operation that require written policies and procedures are not possible. Clearly, the policy and procedures of a large, multinational restaurant chain will be very different from the needs of a small, independently owned sandwich shop.

Despite differences in size and need, however, all foodservice operations undertake worker-related activities that can be readily identified. There are a variety of ways to classify these activities and the policy- and procedure-making related to them. Figure 4.2 lists one good way to categorize the areas of policy and procedure responsibility common to all foodservice operators. It categorizes the areas of worker-focused policy and procedure development as related to

✓ Staffing the organization
✓ Developing staff
✓ Motivating staff
✓ Maintaining staff

Assigned Area/Activity	Requires Policies and Procedures Related to
Staffing the organization	Employee planning and needs analysis
	Recruiting
	Interviewing
	Selecting
Developing staff	Employee onboarding and orientation
	Training
	Employee development and career planning
	Managing organizational change
Motivating staff	Job design
	Employee evaluation
	Compensation
	Employee benefits
	Employee recognition
Maintaining staff	Employee health
	Employee safety
	Employee security
	Employee-related communication

Figure 4.2 Policy and Procedure Development Areas and Activities

Staffing the Organization

The recruitment and selection of employees (see Chapters 5 and 6) is likely the area that most operators think of first when they consider staffing their businesses. Before recruiting and selecting employees, however, operators must carefully assess their staffing needs.

If, for example, an operation's recipes are complex and difficult to execute consistently, then highly skilled food production staff must be secured. If, alternatively, the menu items to be served are rather simple to prepare, extensive culinary skills will not be necessary for production staff.

In all foodservice operations, before employees can be recruited, their skill requirements must be established. As a result, even at the smallest operations, the foodservice operator must identify any specific skills, knowledge, and abilities of the needed employees. In addition, it is important to recall that the specific requirements of current labor law mandate that operators thoroughly understand the specific skills required for the jobs they advertise. Identifying and documenting those specific skill sets effectively helps limit the potential legal liability that could be incurred if protected classes (see Chapter 2) of employees are excluded from the search process.

Excluding potential employees on the basis of identified and legitimately required job skills is always legal. Excluding potential candidates for non-job-related reasons is most often illegal. When the critical characteristics related to a job's successful candidates have been carefully identified, the two most important staffing-related tasks facing a foodservice operator are:

1) Ensuring an adequate pool of qualified applicants to maximize the operation's chances to hire an outstanding candidate.
2) Providing sufficient job information to discourage unqualified job applicants so the operation will not waste the time and resources required to interview non-qualified personnel.

Notes: The policies and procedures related to employee recruitment and selection are among some of the most important for any foodservice operation. The next chapter of this book describes in detail how operators develop policies and procedures to ensure fair hiring practices are used.

After an adequate number of qualified candidates are identified, it is the operator's job to make a hiring decision (in a small operation) or to refer those candidates to the individual who will make the hiring decision (in larger operations). In either case, testing and/or other assessment steps may precede the actual job offer.

Identifying qualified candidates and offering positions to them is only a part of the operator's job. In a tight labor market, qualified and talented applicants are likely to be sought by a variety of competitors both within and external to the foodservice industry. Therefore, the operator must also encourage the desired candidate to ultimately accept the position.

To help do this, the operator might typically provide the candidate with a good deal of job and organization-related information. Topics such as organizational culture, growth plans, and performance expectations are all notable areas that could influence an individual's acceptance decision, and these should be fully discussed with the candidate. Information related to these subjects should be accurate and help the candidate make an appropriate career decision that is best for the candidate and the foodservice operation.

In nearly all cases, workers who accept a new position with a foodservice operator do so with the hope that they will like their work and the organization that employs them. Operators can help employees make the right job choice when they are honest and realistic about each new employee's job and the employee's future potential with the business. Establishing unrealistic expectations for the employee will benefit neither the new employee nor the foodservice operator.

There are some questions nearly every new employee would like to have honestly addressed by their potential employer before they accept a job. These include:

✓ How secure is my position?
✓ What will be the financial and nonfinancial benefits associated with performing my job well?
✓ Are there realistic possibilities for promotion? What are they?
✓ Who should I talk to during my first few weeks of employment if I am having trouble adjusting to the new job?

It is certainly in the short-term best interests of foodservice operators to fill their vacant positions. However, in the long run, it is in their best interests to fill vacant positions with employees who have a realistic understanding of the limitations and potential available to them in their new jobs. Doing this can help in reducing employee turnover rates and in enhancing their employees' job satisfaction levels. This requires operators to be both informative and honest as they describe positions in their organizations to all potential employees.

Developing Staff

After new employees are selected, properly orienting them to the organization becomes an important task. Details about how this is best done are provided in Chapter 7 of this book. However, it is important to recognize that even very experienced employees who need little or virtually no skills training will still need to learn much about their new employer.

Information about items such as an operation's rules, regulations, and goals need to be communicated. Procedurally, questions of who will do the orientation, when it will occur, and what specific topics will be addressed are all employee-related policy and/or procedure issues.

In some cases, employees may be qualified for the job they have secured but will still require facility-specific skill training. For example, even a baker with many years of experience will likely need to be shown "where we store it here" and "how we do it here" when they begin working for a new foodservice operation. Minor variations in baking procedures such as the preferred size of dinner rolls, or fillings to be used for pastries must be taught. Similarly, even experienced service staff, if newly hired, will likely need to be instructed on a restaurant's specific table settings, order taking, order entry system, order pick-up, food delivery, and check presentation procedures.

As an employee's career within a foodservice operation progresses, that employee may need to acquire new skills. In many cases, changes in the employee's work tasks and/or in the goals and needs of the operation may dictate that additional training is needed.

It is also important to remember that many employees hope to advance within the organization that employs them. The foodservice operator should provide those workers, to the greatest degree possible, with opportunities to do so. This may take the form of providing applicable employees with advanced skills training related to their present jobs, training in jobs they may hold in the future, or **cross-training** employees in new skills to prepare them for different jobs.

The best foodservice operators know that planning for their future staffing needs is an ongoing process. The competitive nature of the foodservice industry requires that most operations have the ability to rapidly add products or services that will directly impact their employees. Newly added menu items, for example, will likely require additional food production skills training. Adding the service of free wireless Internet to a coffee shop may require that one or more employees receive additional training in computer-related technology. Regardless of the individual within the organization who actually does the training, it remains the operator's job to ensure that proper training is provided.

Key Term

Cross-training: A training tactic that allows employees to learn how to perform tasks in another position.

Motivating Staff

The task of motivating employees to do their best is one of the most studied, talked about, and debated of all employee-related topics. The question of how to motivate employees to do their best (or even if it is possible for operators to do so) will continue to be an important topic. However, one helpful way to consider the role of foodservice operators as they develop policies and procedures related to employee motivation is to consider two factors that are commonly agreed to affect worker motivation. These are an employee's

✓ Ability to do a job
✓ Willingness to do a job

The ability of a worker to effectively do a job is affected by the employee's skill level, the availability of effective training, and the worker's access to the tools or information needed to properly complete assigned tasks.

The willingness of employees to work efficiently and effectively has long been the subject of study by motivational theorists and managers. Perhaps the most important thing for foodservice operators to recognize related to employee motivation, however, is that different groups of workers are most often motivated by different factors.

For example, a college student working part-time at an operation during the summer months to fund their next semester's tuition will likely consider the amount of money they earn to be a major motivational factor. Alternatively, a healthy retiree working in the same operation may be motivated primarily by an interest in staying active, and perhaps simply enjoying their interaction with guests and their co-workers. A foodservice operation's work policies and procedures should be designed to appeal to workers with a variety of motivational interests.

Find Out More

Much of the published literature and traditionally communicated theory related to employee motivation tends to focus on professional or "salaried" workers. While it is certainly important to keep these types of workers motivated, in the foodservice industry the great majority of workers are paid hourly.

In addition, many of these workers are part-time employees and often they may work for more than one employer at the same time. As a result, the factors that motivate hourly paid foodservice employees are unique.

Some of the most important motivation and retention-related factors for part-time and hourly paid workers in the foodservice industry include:

1) Timely communication of work schedules
2) Flexibility of work schedules
3) Perceived fairness of work schedules
4) Being considered a respected member of the operation's team
5) Participating in a fair tipping system (for tipped employees)
6) Being involved in decision-making when possible
7) Reasonable break and meal policies
8) Safe working conditions
9) Respectful supervisors
10) Being supported when guests make inappropriate comments to them or behave inappropriately toward them

To review some additional helpful suggestions for attracting and retaining part-time and hourly paid staff, enter "Tips for motivating hourly paid restaurant employees" in your favorite search engine and review the results.

Maintaining Staff

Even the best of work teams requires regular maintenance and care. Policies and procedures related to the maintenance of employees include those that help encourage quality workers to stay with the organization. Major areas of concern include worker health and safety, and the development and implementation of employee assistance programs (see Chapter 3).

Additional areas of staff maintenance relate to communication efforts designed to keep employees informed about the work-related issues that are important to them. Other policies may identify opportunities for employees to have their voices heard by management. Staff meetings, bulletin boards, newsletters, mass e-mails, and suggestion boxes are common examples of processes foodservice operators can routinely use to encourage information exchange.

Steps in Employee Policy and Procedure Development

Foodservice operators must be thoughtful as they develop and implement their policies and procedures. In most cases, making worker-related decisions based on momentary operational situations is a poor practice. To illustrate, consider the dining room manager who, because he was rushed and harried during a busy dinner period, "fires" an employee (a busser) who (the manager feels) violated the dress code because the busser's shirt was untucked while clearing dining room tables.

In this case, the busser's shirt was indeed untucked. Therefore, the dining room manager's actions might seem reasonable. Experienced foodservice operators advising this dining room manager, however, would likely first ask a few relevant questions.

1) Was the employee ever informed of the operation's "no untucked shirts" policy? Is that requirement specifically listed within the operation's dress code?
2) Is there *written* evidence that the employee received and understood (in their own language) this specific policy?
3) How long has the employee been with the organization?
4) Is this the employee's first dress code policy violation?
5) Was the employee given a reasonable chance to explain the circumstances leading to his violation of the policy?
6) In the past have all employees found to be in similar violation of this same policy been fired immediately? If not, what was the rationale for firing the employee in this example?
7) Under the laws of the state in which the restaurant is located, will the employee likely qualify to receive unemployment compensation in this case?
8) In the past, has this manager been rightly (or wrongly) accused of discrimination in the hiring or firing of the operation's employees?
9) Given the circumstances of this incident, what message did the dining room manager seek to send to the operation's remaining employees?
10) Does the manager feel that the message sent will help or harm the restaurant's long-term worker-relations efforts?

As this example illustrates, foodservice operators are generally given broad powers to hire and terminate employees. However, today's legal environment and the basic concept of fairness and quality worker relations mandates that operators carefully follow policies that have been thoughtfully developed when managing their employees.

To help minimize the negative consequences that can be associated with improperly developing or applying an operation's policies and procedures, thoughtful operators should establish a basic policy and procedure development process. While this process varies based on the size and type of operation involved, most follow a series of six important steps that help ensure an appropriate development approach is used. These steps are the following:

Step 1. Identify the issue to be addressed. Policies and procedures typically are developed to address an important issue, establish a standard, or solve an identifiable problem.

Step 2. Consider on-site factors affecting implementation. In this step, internal factors directly affecting the development of policy or procedures are considered. Examples include items such as the existence of a union contract, the objectives an operator seeks to achieve, and the time frame required for implementation.

Step 3. Consider off-site factors affecting implementation. Off-site factors that may be considered in the policy and procedure development process may include consistency with an owner's other properties, local, state, and/or national labor-related legislation, and direct competitors' policies.

Step 4: Draft policy and procedures and submit them for legal review. After a policy and the procedures required to implement it have been drafted, it is always a good idea to have the draft examined by a qualified legal expert. This step is important in helping to reduce potential litigation directly related to the policy and is addressed in detail in the next section of this chapter.

Step 5: Develop related documentation and recordkeeping requirements. After a legal review is completed, operators develop the recordkeeping procedures needed to ensure the consistent application of the policy and provide the proof that it will indeed be applied consistently. This is done via documenting the training of those staff members who will enforce the policies.

Step 6: Communicate finalized policy and procedures to affected parties. Policies and procedures that have not been adequately communicated to those affected are difficult or perhaps impossible to enforce. The final step in policy development and implementation is the policy's clear and timely communication to all affected parties (employees) and the documentation of that communication.

What Would You Do? 4.1

"I don't think we have a choice," said Jona, the assistant manager at the Brandywine Bar and Grill. "With our current policy, it's just too hard to find enough staff."

Jona was talking to Anna, the Brandywine's owner, and general manager.

"I just read an article that said over 35% of Americans between the age of 18 and 29 have at least one tattoo," continued Jona, "and lots of them have more than one. Today I interviewed a young lady who had a really cute butterfly tattoo on the back of her left hand. Unless we required her to wear gloves when she was serving guests, our current 'no visible tattoos' policy keeps her out of our candidate pool. And I think she would have been a really great employee for us."

"O.K. I hear what you're saying," replied Anna, "but what if it had been a tattoo of a snake wrapping itself around her throat? Would that be OK too? I think we are on a slippery slope if we start making exceptions based on what tattoos we think are nice and which ones we think might offend our guests."

"Well, I think we're going to have an increasingly hard time getting and keeping workers unless we make some exceptions that make reasonable sense!" replied Jona, "we won't have any guests to offend if we don't have any servers available to serve them!"

Assume you were Anna. Do you agree with Jona that it might make sense to consider modifying your current no visible tattoo policy? How important would it be that any new policy you implement be applied in a way that makes good sense to your future potential employees as well as to your current guests?

Reviewing for Legal Compliance

As previously indicated, a thorough legal review is an important step in the policy and procedure development process. The reason this step is critical is readily apparent when operators consider that developing procedures to support an illegal policy makes no sense. While experienced operators understand that the way in which a policy is implemented can be flawed, a policy that is already flawed or illegal from the outset simply should not be implemented.

A legal review of a policy proposed by an operator does not always indicate that the proposed policy is illegal. Rather, the legal review more likely indicates potentially troublesome procedural areas to which an operator should pay close attention.

To illustrate how a legal policy that is improperly applied could create difficulties and, as a result, show the importance of a thorough legal review, consider the case of Nevaeh. She is the foodservice director at a local hospital. Nevaeh's operation prepares and serves more than 300 meals per day. Nevaeh knows that the law allows her significant discretion in setting appearance standards for her staff, and she wishes to do so by creating and implementing a department-wide dress code. *Note*: She has every right to do so.

In nearly all cases, operators such as Nevaeh can (and often do) legally impose rules and guidelines that have a basis in social norms such as those prohibiting visible tattoos, body piercings, or earrings for men. While tattoos and piercings may be examples of employee self-expression, they generally are not recognized as signs of religious or racial expression (and therefore are not typically protected under federal discrimination laws).

For example, in *Cloutier v. Costco Wholesale Corp.* (390 F.3d 126 [1st Cir. 2004]), the First Circuit Court considered whether an employer was required to exempt a cashier from its dress code policy prohibiting facial jewelry (except earrings) and allow her to wear facial piercings as a reasonable religious accommodation. The employee claimed that her religious practice as a member of the Church of Body Modification required that she wear the piercings uncovered at all times.

This court accepted that the cashier was protected by Title VII of the Civil Rights Act, without specifically discussing the sincerity of her beliefs, and ruled only on whether her requested exemption from the dress code would impose an undue hardship on the employer. The court found that exempting the employee from the policy would in fact create an undue hardship for the employer because it would "adversely affect the employer's public image," and the employer had a legitimate business interest in cultivating a professional image. The employee's case was dismissed. Foodservice operators such as Nevaeh have the same legitimate business interest.

In most cases, a carefully drafted dress code that is applied consistently does not violate discrimination laws. Despite the wide latitude given to foodservice operators, however, all or part of an implemented dress code may be found to be discriminatory. It is not uncommon for foodservice workers to challenge even well-designed dress codes based on a purported discrimination related to their sex, race, and/or religion.

Sex discrimination claims related to dress codes are not usually successful unless the dress policy has no basis in social customs, differentiates markedly between men and women, or imposes a burden on women that is not imposed on men. For example, a policy that requires female foodservice operators to wear uniforms while male foodservice operators are allowed to wear "professional attire," such as their choice of suit and tie, is likely discriminatory.

However, dress requirements that reflect current social norms generally are upheld, even when they affect only one sex. For example, in a decision by the Eleventh Circuit Court of Appeals in *Harper v. Blockbuster Entertainment Corp.* (139 F.3d 1385 [11th Cir. 1998]), the court upheld an employer's policy that required only male employees to cut their long hair.

In most cases, race discrimination claims would be difficult for an employee to prove because the employee must show that the employer's dress code has an unfair impact on a protected class of employees. One dress code-related area where race claims have had some success is in challenges to "no beard" and "acceptable hair style" policies.

A few courts have determined that a policy that requires all male employees to be clean-shaven may discriminate if it does not accommodate individuals with *pseudo folliculitis barbae*, a skin condition aggravated by shaving that occurs almost exclusively among African-American men. Similarly, in 2023, the State of Michigan amended an existing law prohibiting discrimination associated with race, including, but not limited to, "hair texture and protective hairstyles," with the bill's sponsor stating, "hair discrimination is nothing more than thinly veiled racial discrimination."[1]

Employees have had the most success challenging dress codes on the basis that the codes violate religious discrimination laws. These charges occur most frequently when an employer is unwilling to allow an employee's religious dress or appearance. For example, a policy may be discriminatory if it does not accommodate an employee's religious need to cover his or her head or wear a beard, or to wear a hijab. However, if an employer can show that the accommodation would be an undue hardship (such as if the employee's dress or grooming created a safety concern) it likely would not be required to vary its policy. Interestingly, dress code claims also may be filed under the National Labor Relations Act (NLRA) (see Chapter 2).

Therefore, even in an area such as dress codes where employers such as Nevaeh have wide latitude to manage their operations as they see fit, the potential for legal difficulties can still exist. Virtually any of the areas in which policies and procedures are developed may be the source of litigation, but operators must be most careful in areas related to the control of employee dress, expression of opinion, and behavior away from the worksite.

To cite just one example of the regulation of behavior away from the website, consider that as recently as 15 or 20 years ago few foodservice operators had any policies or procedures regulating an employee's postings on social media sites. Today, of course, posting on and communicating with others via Facebook, YouTube, X (formerly Twitter), WhatsApp, Instagram, and other similar electronic venues is extremely popular, and especially so with younger workers.

Most of a worker's posts and tweets will be of a personal nature; however, some inevitably will have to do with work. Just as inevitably, the comments workers make about their workplace, or their immediate supervisors and guests are

---

1 https://www.lansingstatejournal.com/story/news/politics/2023/06/15/michigan-civil-rights-law-hair-discrimination-ban/70307716007/. Retrieved June 15, 2023.

sometimes less than complimentary. It may be tempting for operators who become aware of derogatory comments to punish those who made them (especially if the statements are unfair or one-sided). However, these operators must tread carefully.

As a result of "lawful conduct" statutes, employees generally have rights to engage in lawful activities during nonwork hours. Communicating electronically is certainly not illegal. In fact, the NLRA grants employees free speech rights that include criticism of their companies and their managers. Furthermore, the NLRA generally protects from retaliation measures those workers who (even publicly) are critical of management. It is just as important to note, however, that not all work-related speech is protected either by the NLRA or laws related to free speech.

Reckless or malicious lies, hate speech, disclosure of confidential company information, and threatening and harassing statements are not typically protected forms of free speech. As a result, savvy foodservice operators recognize they must carefully navigate the sometimes very fine line between their companies' rights and the free speech rights of their employees. In this example, a legal review of any policies addressing employee's rights to communicate freely in electronic formats is certainly beneficial.

Experienced foodservice operators recognize that a periodic legal review conducted by a legal professional of all worker-related policies and procedures, and a specific review whenever they are significantly modified or revised, is a wise use of time and money.

Find Out More

Cannabis use in the United States has a long and complicated history. Currently the use, sale, and possession of cannabis are illegal under federal law.

Many (but not all!) states, however, have enacted their own laws that are contradictory to federal law. In some states, the use of medical marijuana is allowed but recreational marijuana is not. In other states, both medical and recreational marijuana use is allowed.

Some states with medical marijuana laws have also instituted protections for a worker's off-duty use with a valid prescription as long as they do not arrive at work under the influence of marijuana. Other states with medical marijuana laws either specifically allow employers to fire employees for off-duty use or do not very clearly address the issue.

For foodservice operators concerned about workplace safety, and especially those who implement pre-employment or post-employment drug testing policies, understanding the laws that relate to workers' marijuana use in their own state is incredibly important.

To learn more about the marijuana-related worker protections (if any) implemented in a specific state enter "Workplace protection for medical marijuana users in (*insert state name*)" and review the results.

Employee Handbooks

The development of a written **employee handbook** is one of the best ways for a foodservice operation to establish and communicate rules of conduct for its workers and to help ensure a healthy, safe, and secure workplace for all.

When employee-related policies and procedures are put into writing, it makes it easier to address potential problem areas as they arise. An employee handbook includes information all employees will want to know about, as well as information they should know.

For example, all employers will want to know about time off benefits they may be allowed to accrue such as personal leave, sick days, and vacations. The number of hours or days accumulated is important to workers, and it is also important to ensure that all employees are treated equally in this regard. Having a written policy allows operators to point to the appropriate area of the handbook to help explain to workers how they are being credited with earned time off. Similarly, if a foodservice operator has instituted a zero-tolerance harassment policy (see Chapter 3) the employee handbook is the proper place to communicate details of that policy to all employees.

While neither federal nor state law requires the use of an employee handbook, there are a number of important reasons foodservice operators should develop one. These include:

1) *Ensuring equitable treatment for all workers.* When worker-related policies and procedures are put into writing it helps ensure that every worker is treated equally. This is important for workers, but it is also important for operators as well. For example, operators can point to the relevant part of an employee manual if a worker makes a claim of unfair or biased treatment.

2) *Fulfilling legal requirements.* Both federal and state employment laws and regulations require foodservice operators to provide specific information to employees to stay in compliance with some employment-related laws. For example, all employees must be informed of an employer's commitment to provide equal opportunity in all aspects of the employment relationship. This applies to all employees without regard to race, color, religion, sex (including pregnancy, sexual orientation, or gender identity). Current regulations also address national origin, age (40 or older), disability, and genetic information (including family medical history), or any other category protected class identified by federal, state, and local laws.

Key Term

Employee handbook:
A permanent reference guide for employers and employees that contains information about a company, its goals, and its current employment policies and procedures.

Also commonly referred to as an "Employee manual."

Similarly, in some states, all employees must, by law, be informed of an operation's policy regarding harassment. For example, in the state of Maine, employers must:

> *provide a written sexual harassment notice/policy to employees on an annual basis. The notice/policy must include at least the following information:*
> - *The illegality of sexual harassment;*
> - *The definition of sexual harassment under state law;*
> - *A description of sexual harassment utilizing examples;*
> - *The internal complaint process available to the employee;*
> - *The legal recourse and complaint process available through the Maine Human Rights Commission;*
> - *Directions on how to contact the Commission; and*
> - *The protection against retaliation.*[2]

3) *Defending against lawsuits.* When a foodservice operation can prove that all employees have been informed about important policies and procedures, the operation is in a better position, if necessary, to legally justify actions it has taken to enforce its policies.

4) *Encouraging open communication between workers and management.* When workers are informed of their rights and responsibilities in writing, it becomes easier to discuss issues directly affecting the workers. For example, if employees are specifically told to whom they should report potentially harassing actions, they will feel more comfortable reporting the actions.

While the specific information that should be included in an employee handbook varies by operation, some items are best left out of an employee manual. These include items such as specific equipment operating procedures, detailed work processes, **job descriptions**, and **job specifications**. Items such as these should be excluded so the entire employee handbook does not have to be updated every time these types of items are modified.

While it is important to ensure that the proper content is included in an employee handbook, it is just as important to document that each employee has been given and understands the information in the handbook. This can be done

Key Term

Job description: A statement that outlines the specifics of a particular job or position within a foodservice operation. It provides details about the responsibilities and conditions of the job.

Key Term

Job specification: A listing of the personal characteristics and skills needed to perform the tasks contained in a job description.

2 https://www.lansingstatejournal.com/story/news/politics/2023/06/15/michigan-civil-rights-law-hair-discrimination-ban/70307716007/. Retrieved June 15, 2023.

by requiring employees to sign a document showing they have received the handbook. Figure 4.3 is a sample of a document that can be used to prove employees have been given a chance to read and review their handbook. This signed document should be put in each employee's personal file, and a new one should be executed anytime there have been significant changes to the handbook and new or revised information has been provided to employees.

Acknowledgment of Receipt of Employee Handbook

This Employee Handbook contains important information about *<insert operation's name>*.

I understand that I should ask my supervisor, manager, or the owner about any questions I have that have not been answered in the Handbook. I have entered into my employment relationship with the company voluntarily and understand that there is no specified length of my employment. Either the company or I can terminate the relationship at will, at any time, with or without cause, and with or without advance notice.

The information, policies, procedures, and benefits described in this manual are subject to change at any time, I acknowledge that revisions to the handbook may occur. All such changes will be communicated through an official notice, and I understand that revised information may add to, modify, or eliminate the company's existing policies.

Also, I understand that this handbook is neither a contract of employment nor a legally binding agreement. I have been given the time needed to read the handbook, and I agree to accept the terms within it. I also understand that it is my responsibility to comply with the policies contained in this handbook and any revisions made to it.

I understand that I am expected to read the entire handbook. After I have done so, I will sign two original copies of this Acknowledgment of Receipt, retain one copy for myself, and return one copy to the company's representative listed below.

I understand that this form will be retained in my personal file.

_____   _____
Employee's Signature   Date

_____
Employee's Name (printed)

_____   _____
Company Representative   Date

Figure 4.3   Sample Employee Handbook Signature Page

> **Technology at Work**
>
> Some operators may feel that the development of an employee handbook for their own operation would be a difficult and time-consuming task. That may have been true in the past, however, today foodservice operators can take advantage of the many employee manual templates available online. These templates may be modified to include the specific policies and procedures selected by an individual operator.
>
> In some cases, the templates are offered for free, or for very little cost. In addition, the best of the entities providing these templates allow an operator to choose the state in which they are located, helping to ensure that any state laws affecting their employee manuals are addressed.
>
> Regardless of the source of the employee handbook template, it is important to emphasize that a completed employee handbook should go through a formal legal review before it is put in place. Doing so helps to protect the operator as well as the operation's employees.
>
> To learn more about templates available for the production of an employee handbook, enter "sample employee handbooks for restaurants" in your favorite search engine and view the results.

Recordkeeping Requirements

In the fairly recent past, most worker-related records in the foodservice industry simply consisted of hard-copy (paper) files stored in the appropriate employee **personal file** or in a file developed specifically for recordkeeping purposes.

For example, information related to employees' requests for time off or paid vacation might be kept in the individual employee's file or in files designed to track and record these types of employee requests. Obviously, in very large foodservice operations with hundreds of employees, such a paper-based system could easily become unwieldy and cumbersome. Increasingly, even in smaller foodservice operations, the management

Key Term

Personal file: An individual worker's file that includes the worker's application for employment, and records which are used or have been used to determine the worker's qualifications for promotion, compensation, and/or termination. Information about disciplinary actions, if any, are also included in the personal file.

of today's worker-related records and information requires the application of advanced technology hardware and software, in part because of increases in recordkeeping requirements and the challenges of maintaining the accessibility and security of worker-related information.

The types of records that foodservice operators must maintain on their employers include both those related to policies and procedures the operator has implemented, and records or documents required by various government and regulatory entities.

Typical records that are likely to be maintained in an individual employee's personal file include:

✓ Employment applications
✓ Resumes
✓ Receipt of employee handbook documentation
✓ Performance evaluations
✓ Disciplinary records
✓ Medical files
✓ Insurance-related records and correspondence
✓ Training records and documentation
✓ Certificates, transcripts, and diplomas
✓ Lawsuit-related information
✓ Other employee or employment-related correspondence
✓ Governmental entity required records

Federal law does not require that an employee be given access to their personal file, but some states do make that requirement. Since a personal file often includes information about a worker's performance, medical issues, and investigations of charged misconduct, these files are generally considered private and should be accessed only by the employer and the employee.

Other worker-related files and documents must be retained by employers. To cite just one example of a governmental entity required record, consider that all U.S. employers must properly complete a **Form I-9** (see Chapter 6) for everyone they hire. This includes citizens and non-citizens, and both employees and employers must complete the form.

On a Form I-9, an employee must attest to their employment authorization. The employee must also present their employer with acceptable documents as evidence of identity and employment authorization. The employer must examine these

Key Term

Form I-9: Formally the "Employment Eligibility Verification," this document is a United States Citizen and Immigration Services form required for use by all employers. It is used to verify the identity and legal authorization to work of all paid employees.

documents to determine whether they reasonably appear to be genuine and relate to the employee, then record the document information on the employee's Form I-9.

Employers must retain Form I-9 for a designated period of time and make it available for inspection by authorized government officers. Failure to have a Form I-9 in the personal file of each worker employed subjects a foodservice operator to penalties and fines. Under current immigration law, I-9 forms should be retained for three years after the date of hire, or one year after the date employment ends (whichever is later).

As foodservice operators make decisions about the worker-related records they will keep on file and for the length of time they will keep them, it is important to recognize that there are both practical and legal implications to these determinations.

To illustrate, consider the case of David Berger who owns and operates a subshop in a strip shopping mall. David has determined that it is a good idea to keep a copy of all employee applications he has received when he advertises to add a new person to his staff.

The reasons David considered before making this determination are numerous but include the ability to monitor the quality of the available workforce. Two other reasons are to help ensure that vacancy advertisements appeal to the broadest labor pool possible, and to help judge the workforce demand for his vacancies. Even after making this decision, however, David has even more things to consider regarding his applicant recordkeeping system including:

1) Are all individuals who submit resumes via the Internet considered applicants for recordkeeping and reporting purposes?
2) If so, how will these records be stored and for how long?
3) What about those candidates who, for a variety of reasons, were clearly unqualified for the position?

Questions such as these may be difficult to address, but David must still do so thoughtfully. Other specific examples of records-related questions that require policy and procedural decisions include:

✓ How do regulations related to specific laws, such as the ADA, FMLA, and FLSA (see Chapter 2), affect the length of time records should be kept?
✓ Should employees have access to their discipline records? Can they make copies?
✓ How long should a terminated employee's files be kept?
✓ Which personnel management–related documents should require actual (not electronic) signatures and/or hard copies?

Regardless of how David answers questions such as these, he must make significant decisions regarding the employment records that must be retained and the length of time to retain them. In some cases, employment-related legislation at the federal or state level dictates the full or partial answers to some questions of this type.

Technology at Work

Policy and procedure-related recordkeeping requirements for foodservice operators can be confusing because there are numerous regulations that govern some aspect of employer recordkeeping and retention. A number of Federal government agencies have their own recordkeeping requirements and for many operators individual state and local statutes and regulations must also be considered.

To complicate matters even further, it is important to know that some recordkeeping provisions apply to most all employers, but many of the obligations are dependent on the number of workers a foodservice operator employs.

The complexity of meeting recordkeeping requirements encourage many foodservice operators to utilize software designed specifically to ensure that they are in compliance with recordkeeping in their own state.

These software programs help to achieve three primary goals:

1) Assist in regulatory compliance.
2) Protect sensitive employee information.
3) Ensure needed records are readily accessible when needed.

To review some employee records-related software designed specifically to assist foodservice operators, enter "employee recordkeeping software for restaurants" in your favorite search engine and view the results.

Experienced foodservice operators agree that determining exactly which employment-related records to keep, and for how long, is one of their most essential tasks. Most would also agree that having well-documented employee-related policies and procedures in place, despite the enormity of the task, is critical for foodservice operators of all sizes.

Foodservice operators who understand their role as team leaders, as well as their legal obligations as employers, are in an excellent position to establish and enforce work policies and procedures that will help them achieve their operational goals. To achieve those goals, however, operators will need a dedicated team of professionals to help them serve their guests in a way that makes the guests want to come back. Recruiting team members who have, or can be taught, the skills needed to reach all of an operation's financial and service-related goals is so important it will be the sole topic of the next chapter.

What Would You Do? 4.2

"But Lonnie, I just can't work on Sunday," said Shingi, a part-time server at Captain Jack's Seafood Shack.

Lonnie, the dining room supervisor at Captain Jack's, had just told Shingi that a large group had just booked a private party for brunch this coming Sunday. The result was an additional 90 guests that had not been included in the original sales forecast when Lonnie had made and distributed this week's dining room server work schedule. So, he had to make a scheduling change that affected Shingi.

"Why can't you work on Sunday?" Lonnie asked.

Then, before Shingi could answer, he added, "You know when we hired you, we told you that our business can be unpredictable and that all employees' work schedules are subject to change. It's clearly stated in the employee manual, and when you signed it, you said you understood all the policies in the manual!"

"I know I did," replied Shingi, "and I'd work if I could, Lonnie, because I really need this job. But, after you posted the original schedule last week and I saw I was off Sunday, I told my boss it was okay to schedule me at my other job. They always work around my schedule here, and I promised them I would be there. Since I already promised them I would come in I think I'll get fired from there if I don't show up. And I need that job too. I'm sorry, I just can't work Sunday."

Assume you were Lonnie. Many employees in the hospitality industry hold more than one job. What are some reasons why they do so? Do you think this situation warrants the future development of a "second job" (or moonlighting) policy for Captain Jack's? What would the policy be if it were designed to be fair to your operation and yet attractive to those like Shingi who must work more than one job?

Key Terms

Policy	Employee handbook	Personal file
Procedure	Job description	Form I-9
Cross-training	Job specification	

Operator's 10-Point Tactics for Success Checklist

Evaluate your need for, and the current status of, each of the following operational tactics. For those tactics you think are important, but not yet in place, develop an action plan for its implementation including who will be responsible for the tactic's completion and the target date by which it should be completed.

Tactic	Don't Agree (Not Done)	Agree (Done)	Agree (Not Done)	If Not Done	
				Who Is Responsible?	Target Completion Date
1) Operator recognizes the difference between an operating policy and an operating procedure.	____	____	____		
2) Operator understands the importance of worker-focused policy and procedure development related to staffing an organization.	____	____	____		
3) Operator understands the importance of worker-focused policy and procedure development related to developing staff.	____	____	____		
4) Operator understands the importance of worker-focused policy and procedure development related to motivating staff.	____	____	____		
5) Operator understands the importance of worker-focused policy and procedure development related to maintaining staff.	____	____	____		
6) Operator can summarize the importance of each of the six steps used to develop effective policies and procedures.	____	____	____		

				If Not Done	
Tactic	Don't Agree (Not Done)	Agree (Done)	Agree (Not Done)	Who Is Responsible?	Target Completion Date
7) Operator can state the reasons worker-related policies and procedures should be subjected to a thorough legal review before their implementation.	———	———	———		
8) Operator recognizes the importance of preparing and disseminating a properly prepared employee handbook to all staff members.	———	———	———		
9) Operator revises their employee handbook as needed and uses a formal process to document employee receipt of the new handbook.	———	———	———		
10) Operator has reviewed the recordkeeping requirements applicable to their business and has implemented a system to maintain all required records for the required period of time.	———	———	———		

5

Recruiting and Retaining Team Members

What You Will Learn

1) The Importance of Effective Job Design
2) How to Successfully Recruit Foodservice Team Members
3) How to Successfully Retain Foodservice Team Members

Operator's Brief

In today's very tight and competitive labor market, obtaining and retaining an efficient workforce is one of a foodservice operator's most significant challenges. In this chapter, you will learn that the identification and assembly of an effective foodservice work team begins with proper job design. How foodservice operators initially design their jobs can have a significant impact on the overall appeal of these jobs. In this chapter, you will learn how jobs can be designed in ways that make them as attractive as possible to your potential employees.

After you have identified and designed the jobs in your operation, you can begin the process of recruiting qualified team members. There are a variety of ways to do this including (1) internal recruiting, (2) traditional external recruiting, (3) online recruiting, and (4) recruiting for diversity.

Internal recruiting addresses the potential of using staff who are currently employed to fill vacancies as they occur. Traditional external recruiting involves the use of signs, posters, flyers, and other means of advertising to attract a pool of qualified applicants. Increasingly, foodservice operators also use online recruiting to expand their pool of potential applicants. Finally, recruiting for diversity is an effective way for foodservice operators to increase the number and quality of potential job applicants, and, as a result, improve the quality of their entire workforce. The best ways to utilize each of these four different and useful approaches are detailed in this chapter.

Foodservice operators can best minimize their future needs to recruit new employees as they retain the high-quality workers who are already employed by them. In this chapter, you will learn specific techniques you can use to help ensure that involuntary loss of your hourly paid employees is minimized and the longevity of your work team is optimized.

CHAPTER OUTLINE

Job Design and Job Creation
    The Impact of Job Design
    Effectively Designing and Creating Jobs
Recruiting Team Members
    Internal Recruiting
    Traditional External Recruitment Methods
    Internet-based External Recruitment Methods
    Recruitment of Diverse Team Members
Retaining Team Members

Job Design and Job Creation

Foodservice operators in all industry segments and all sizes face a variety of challenges as they attempt to achieve operational goals. Consistently, however, the number one challenge foodservice operators report is that of obtaining and maintaining a sufficient number of qualified staff members.[1]

To achieve their goals, all foodservice operators must recruit and retain a team of dedicated professionals to help them properly serve their guests. Experienced foodservice operators recognize, however, that employee recruitment and retention efforts must begin far in advance of advertising their position vacancies.

To illustrate, consider the situation facing Jeff and Nancy Gardner, a married couple who have decided to open their own 100-seat restaurant. It will feature Florida Gulf Coast specialties including soups and chowders, grilled and blackened fish items, and Southern style desserts. Jeff and Nancy know they will need talented front- and back-of-house staff if their operation is to be successful.

Before they can begin to recruit their employees, however, they must have a solid understanding of what the employees will do in each position, and how they

---

1 https://orders.co/blog/challenges-restaurant-face-and-how-to-solve-them/. Retrieved May 5, 2023.

will do it. To begin the process, they must create a **task list** for every position in their new operation.

From their position task lists, it will be possible for them to create a **position analysis**.

A position analysis identifies tasks that will be included in job descriptions (see Chapter 4) which in turn are used for employee recruitment. Therefore, a position analysis should be developed before the job description and the job specification (the listing of the personal characteristics and skills needed to perform the tasks contained in a job description) are prepared.

A position analysis will also be essential as Jeff and Nancy establish **performance standards** for each of their jobs.

Performance standards are essential to the long-term retention of workers. It is important to note that it will not be sufficient for Jeff and Nancy to recruit and hire qualified employees. In addition, they must also retain those employees who meet their jobs' performance standards.

Key Term

Task list: A detailed list of all tasks in a position.

Key Term

Position analysis: A process that examines each task listed for a position and explains how it should be done, with a focus on required knowledge and skills.

Key Term

Performance standards: Measurable quality and/or quantity indicators that identify when an employee is performing a task correctly.

Performance standards should specify required quality and quantity outputs for each task in a task list. For example, the proper quality of a restaurant's dessert is that which is expected when the applicable standardized recipe is followed. The quantity of work output expected of a front-of-house server may be the number of guests properly serviced during the server's shift.

To effectively recruit and retain employees for their restaurant, it is important that the performance standards established by Jeff and Nancy be very clearly defined. When this is done, potential employees know what is expected of them and when their performance is acceptable. Performance standards should be reasonable (challenging but achievable).

As Jeff and Nancy develop their new employee training programs (see Chapter 8), it is essential that all new employees be given the information, tools, and equipment needed to enable them to achieve the performance standards established for their jobs.

The Impact of Job Design

To illustrate the importance of job design and job creation on future employee recruitment and retention successes, consider that Jeff and Nancy must hire personnel who will perform some cleaning tasks. For example, tabletops, chairs, carpets, and windows will likely need to be cleaned after each major meal period. In many foodservice operations, these and related tasks are assigned to front-of-house service staff. Therefore, Jeff and Nancy must decide whether their service staff positions will include these cleaning tasks, or if these cleaning tasks will be assigned to other back-of-house janitorial staff.

The recruiting impact of this decision is clear. Most service staff are tipped employees. In addition, many service staff are paid a rate lower than minimum wage because they allow operators to apply a **tip credit** to the servers' pay.

Assume, in this example, that Jeff and Nancy determined that two hours of cleaning between shifts is typically necessary. If this work is assigned to one or more servers, the servers will be paid less than the minimum wage to complete the cleaning tasks. While this might initially appear financially beneficial to Jeff and Nancy because of the tip credit they can use, also consider the situation that confronts them as they recruit servers.

Key Term

Tip credit: A legally permissible system that allows employers to include tips as part of a worker's wage calculation. Employers can credit a portion of an employee's received tips towards their minimum wage payment obligations so that employers do not have to pay the full minimum wage.

For example, consider an operation that competes with Jeff and Nancy that determine cleaning will be done by the back-of-house janitorial staff, and servers will only be scheduled when they can earn tips. In this situation, it would not be a difficult decision for servers to determine they are better off deciding to work for the owners and operators who do *not* assign cleaning tasks.

To cite a back-of-house job design example, Jeff and Nancy must also hire qualified staff to produce their menu items. Some of the items that will require advanced culinary skills include production of cream sauces, boning fresh fish, and char-broiling oysters. *Note*: Other kitchen tasks such as cleaning vegetables and fruits or chopping greens for salads are less complex.

If Jeff and Nancy assign relatively low-level food production tasks to the same workers who can perform higher-level tasks, two results are likely. First, Jeff and Nancy will be paying higher wages than needed because overly skilled workers will be doing some routine kitchen tasks. The second (and more important result) is the likelihood that highly skilled culinarians will be less interested in doing routine food preparation tasks than tasks that are more challenging. Therefore, if a highly skilled culinary worker considered a job offered by Jeff and Nancy compared to a job offered by a competitor that provided the same rate of pay but more interesting work, the potential candidate would not likely decide to work for Jeff and Nancy.

Effectively Designing and Creating Jobs

As they design and create jobs, foodservice operators in the United States (and in many other parts of the world), must recognize that, for many workers, *what they do* is often seen as a powerful statement about *who they are*. This is not necessarily bad or good, but the result is that foodservice workers are often seen by others to be bartenders, servers, bussers, cooks, or dishwashers.

In some cases, the jobs' titles make it difficult for lower-paid workers to gain high levels of self-esteem by using what they do to define who they are. However, everyone who performs tasks in a job wants to feel like they make a real difference in the lives of their families and others. The foodservice industry employs many highly paid individuals, but some foodservice employees earn low rates of pay and do not readily see high levels of status in their own jobs.

It is important to realize that many foodservice workers do not believe their jobs automatically provide them with a good source of motivation. This can occur even though they have a need to feel good about themselves and most importantly, about what they do each day at work. Effective foodservice operators recognize and encourage the importance of a sense of purpose for the work their team members do. For owners/operators themselves, a sense of purpose likely comes from conceiving and then achieving the vision they have created for their operations. They must also remember that their lower- and minimum-wage employees seek that same sense of purpose.

Foodservice operators must recognize the value of their team members to those they serve and then communicate clearly to those employees about the importance of their contributions. Addressing this key need to affirm the value of what they do and who they serve is an activity that can and should be undertaken by operators for the benefit of their workers and their businesses.

This can be done when operators design jobs that enable the maximum satisfaction possible for their workers. Doing this results in a more highly motivated staff and a more pleasant workplace. It also makes it easier for operators to recruit and retain the workers they need to achieve their operating goals.

It is the authors' contention that, in some cases, the work attitudes of Americans have led directly to poor performance in the designing of many foodservice jobs. To illustrate why, consider what one winner of the popular American television show, "American Idol," said while speaking to the public following her triumph:

> *If you work hard enough, and don't give up, you can reach your dreams just like I did!*

That sounds very good, but the reality is that, in the year she won the competition over 100,000 other contestants also tried to reach their dream of winning along with her but failed to do so. Many likely wanted it just as much. Many likely tried just as hard, but there are numerous factors involved in reaching the top in any field, and the space is limited. Is it always possible to get there with hard work alone? Usually not.

In 2021, 76.1 million workers aged 16 and older in the United States were paid hourly rates, and this represented 55.8% of all wage and salary workers. Nearly 52 million U.S. workers—32% of the country's workforce—earned less than $15 an hour.[2] The great majority of these lower-wage jobs lack benefits such as health insurance or retirement accounts, and, in many cases, they provide little or no chance for career advancement. These conditions translate into millions of Americans who earn near-poverty level incomes, while millions more struggle to make ends meet. Many of these workers are employed in the foodservice industry.

Given the uniquely American view that if people are simply industrious enough, they will inevitably be successful in life leaves many to conclude therefore that being a low-wage earner means a person must be lazy, unskilled, unmotivated, or something similar, and, because of that, they should actually consider themselves to be "lucky" to be offered a job of any type!

The reality is that most low-wage earners are often anything but "lucky" economically. Many of these persons work harder physically than those in more skilled positions must work. Many do not have the time, or the financial resources needed to acquire more education or training. They often are not eligible for significant career advancement, increased benefits, or wages no matter how hard they work or how effectively they perform their jobs. As well, many of these individuals work two or more jobs just to "get by," and they are not lazy, but they do deserve a break and an opportunity!

Caring foodservice operators utilizing thoughtful job design can be the professionals who provide this opportunity, and in doing so help employees and the operator's long-term recruitment and retention efforts.

Foodservice operators facing staffing challenges can help themselves be more competitive as they initially design their jobs. Key areas of concern to operators as they design their jobs in ways that will make them competitive in the labor market include:

✓ Compensation
✓ Benefits
✓ Scheduling
✓ Working conditions

Compensation

The compensation required to be competitive for recruiting and retaining workers can vary greatly based on a foodservice operation's physical location.

---

2 https://www.bloomberg.com/news/articles/2022-03-22/federal-minimum-wage-1-in-3-us-workers-make-less-than-15-an-hour#xj4y7vzkg. Retrieved May 5, 2023.

Those businesses located in high cost of living areas typically must offer greater pay for the same work than do businesses located in lower cost of living areas. Similarly, in those areas where the supply of labor is severely limited, operators will find that higher wages are required to attract a sufficient number of workers.

It is also true, however, that an employer need not offer the *highest* wages in an area to be considered the area's employer of choice (see Chapter1). Competitive pay is important, but it is often *not the most important factor* in securing talented foodservice workers. To address the issue of appropriate pay, foodservice operators should survey their competitors, consider the economic conditions in their area, and then establish pay rates or a **pay range** for each job title they have established.

Key Term

Pay range: The span between the minimum and maximum hourly rate an organization will pay for a specific job or group of jobs.

Benefits

Nearly every foodservice operation offers some level of voluntary benefits to its employees. On the very lowest end of the scale, this may include offering free or reduced meals during working shifts. On the other end of the scale, employers may offer health insurance, paid vacation time, and the matching of 401(k) contributions.

Regardless of the level of benefits offered, it is important for operators to recognize that benefits will be a significant determining factor in many employees' decisions about whether to accept or decline the operator's employment offer. This is especially the case when the pay rates or pay ranges offered for a specific job in an area are relatively similar.

To illustrate, consider the foodservice operator seeking to hire a tipped server at a rate of $10 per hour, the going rate in a specific area. This foodservice operator offers only reduced-cost meals to its employees. A competitive operator, however, may offer the same $10 per hour rate of pay, but also includes a $200 per-month contribution toward a health care plan the business offers and two weeks' paid vacation after one year of service. In this example, it is fairly easy to see that a high-quality server will likely choose to work for the employer offering the greater benefit package. In fact, a high-quality employee may choose to work for the second operator even if the hourly rate of pay offered was only $9.50 per hour!

Scheduling

The work schedules (see Chapter 10) to be assigned to foodservice employees are particularly important in their employment decisions, and this is especially so for those who work part-time. Potential employees can choose to work part-time for a variety of reasons including:

Childcare issues: These workers choose to work part-time for reasons related specifically to affordable, available, and/or adequate childcare.

Personal obligations: These can include family or home-related reasons which may include staying home with a sick child or parent, or to do housework or another chore for the family.

Health or medical issues: The worker's own illness, injury, or disability may prevent him or her from working full time.

School or training: A staff member may work-part time so they can attend high school, college, or another training program.

Retired and/or social security earnings limit: Some employees work part-time because they are retired or because they cannot work more hours without losing a portion of their Social Security benefits.

It is easy to see that in all these cases (and more), the specific work schedule that an employer offers a potential employee can make a great deal of difference to the willingness of the potential employee to accept a job. It is also true that a work schedule deemed acceptable by one part-time employee may not be acceptable to another staff member. Therefore, employers generally do best when they allow maximum scheduling flexibility in the jobs they offer.

Some foodservice operators feel they do not have any flexibility regarding employee scheduling. These operators may maintain that employees are needed when necessary work must be performed, so scheduling flexibility is not an option. However, in many cases these operators are wrong.

To illustrate, consider the operator whose bar area is particularly busy on Friday and Saturday nights. While the operator normally can fully staff the bar with one bartender, on Friday and Saturday nights the operator needs two bartenders. The operator has proposed a shift of 6:00 PM to 2:00 AM, to cover the entirety of the bar's busy time.

If the operator creates a part-time bartender's job consisting of two weekend nights per week at 8 hours per night for a total of 16 hours part-time work, it is likely the operator will have extreme difficulty filling this position.

The reasons are fairly simple. Very few part-time workers will want to work every Friday and Saturday night. Those potential employees who hold full-time Monday through Friday jobs are also not likely to want to spend their entire weekend working. As well, younger workers who might be attracted to this job may find it objectionable to work every Friday and Saturday night. The reason: it would virtually eliminate normal social activities with friends and family.

In this example, it might be more sensible for the operation to provide one or more of its server positions with the training needed to serve as bartender, and then schedule one of those servers on each of the Friday and Saturday nights with busy bartending shifts. In this way, scheduling properly allows the operator to fill their position with qualified staff, which may be extremely hard to otherwise fill.

Working Conditions

For many lower-waged foodservice workers, the pay rates, benefits, and schedules offered by potential employers may be fairly similar. Then working conditions may become the most important factor when selecting one employer over another and in retaining that employer over an extensive time.

If they were asked, most foodservice operators would likely say they want to provide their workers with a "good" place to work. Successful foodservice operators recognize their responsibility to provide a healthy, safe, and secure work environment. They also have created policies and procedures that ensure their employees are as free as possible from risks posed by the workplace, other employees, and guests. However, this recognition is not typically enough to ensure that an operation can attract and retain a sufficient number of necessary workers.

Tips for creating a welcoming and supportive work environment are numerous, and effective foodservice operators can choose from those that apply most directly to them. There are, however, some no-cost characteristics of a positive workplace that all foodservice operators can offer to help ensure the working conditions in their operations are as welcoming as they can be.

Among the most important of these are:

1) *Demonstration of genuine respect.* Respect for all team members is the foundation of a positive work environment. Every foodservice worker deserves to be treated with respect, and every worker has an obligation to treat others in the same way. Demonstrating genuine respect begins with operators and supervisors who foster respect in the workplace through leading by example.

   Regardless of the tasks they are assigned, every foodservice worker wants to know they are valued, appreciated, and will be treated fairly. In many cases, when foodservice workers are asked why they left a previous employer, the answer is not low pay or a dislike of the actual work. Rather, it is frustration resulting from not being treated in the way that the worker felt was most appropriate.

   Improper comments, gestures, and actions can be initiated by management, co-workers, and even guests. It is a major responsibility of foodservice operators who want to ensure their employees elect to join and stay with the organization to regularly confirm the worth of each worker. This can be done as easily as verbally welcoming workers as they arrive at work and thanking them for a job well done as they depart from the job.

2) *Open communication.* Workers in every foodservice operation experience times of stress and frustration. Sometimes this results from the work itself. For example, when a foodservice operation is very busy the demands on servers and production staff can be high. While stress of this type may not be avoidable, it is possible for all employees to feel like they have a voice at work.

In many cases, front-of-house staff deal directly with guests and, as a result, they may have more interaction with the policies and procedures that affect themselves and the guests than do foodservice operators themselves. Therefore, it is important that operators encourage feedback from these employees. **Open-door communication** policies are nearly always present in an operation that recognizes the importance of open communication.

From its definition, one might imply that an open-door communication policy is passive, and that operators must wait for employees to initiate communication. However, that is simply not the case. An open-door policy implies that communication goes both ways and, in many cases, this communication should be initiated by foodservice operators as they ask about their workers' issues and concerns their team members may have. Again, this is a critical yet low-cost technique that requires only a foodservice operator's time and a genuine interest in the concerns of team members.

Key Term

Open-door communication (policy): A policy that indicates to employees that a supervisor or manager is open to discuss an employee's questions, complaints, suggestions, and concerns.

3) *A sense of teamwork.* Few management topics have written about more than the importance of teamwork to an organization's success. Often authored by sports personalities or successful business persons, there is little disagreement with the statement "teamwork makes the dream work."

It is true that teamwork is most often essential to success. However, it is also essential to creating positive work environments that make employees want to select an employer and stay with that employer because they are interested in being part of a cohesive and welcoming work team.

One often neglected aspect of team building involves addressing team members who are not productive and may even be disruptive to the team. Some supervisors may attempt to change these employee's attitudes, and that may be a worthwhile effort in some cases. However, foodservice operators must sometimes recognize that specific team members must be removed from the team. This can be particularly challenging given that worker shortages exist, and terminating an unproductive employee will tend to make that challenge even greater.

However, it is more important to recognize that one improperly behaving team member can make it difficult for many workers to enjoy their jobs. Taking positive action to retain workers as a group is always more important than retaining a single individual who may continually be disruptive to a team.

It is also important to recognize that a positive team atmosphere in the workplace is important for long-term employee retention. However, this will not be immediately apparent when potential employees apply for a job, and some potential employees may not have experienced an operation's sense of teamwork. Therefore, activities that a foodservice operator can undertake to demonstrate a positive work environment during the interview process can be helpful in demonstrating an operation's positive team environment.

Find Out More

Team-building activities including short meetings, external social gatherings, and easy-to-play games are examples of team-building activities foodservice operators can use to build camaraderie among their employees. Team-building activities can be used in every type of foodservice operation. Team-building activities allow workers to get to know each other better as they discover team members' interests, strengths and weaknesses, and how they communicate.

Team-building activities also build camaraderie and trust, which are some of the most important aspects of a successful team. Most importantly, these team-building activities help remind employees that work is never just about them, but rather it is about the entire group.

When workers are encouraged to engage in a group activity rather than work alone, it helps remind team members that a group's success (and ultimately that of the operation as well) should be the priority. This can be particularly useful for teams that struggle with teamwork, are overly competitive, and/or have in the past lacked a feeling of unity.

To learn about no-cost or low-cost creative ideas for enhancing work teams, enter "Team building activities for small businesses" in your favorite search engine and review the results.

What Would You Do? 5.1

"But why are you leaving?" asked Tamara, a day shift server at King's family restaurant. "The tips are just as good here as they will be at the pancake house, and the pay has to be about the same!"

Tamara was talking to Lena, her good friend, and her fellow server at King's. Lena had just confided to Tamara that she had accepted a job starting next week at the Colonial Pancake House, a restaurant that was a direct competitor of King's.

"I know," replied Lena "The pay is the same, but when I talked to them, they really made me feel like they needed me, and that I could make a difference over there. I'm not sure our boss here will even notice that I'm gone. He never says anything to me now."

"Well, I don't know about that," said Tamara, "and he's the same way with me and the other servers. It sure will be different when you are not here. You are one of our best servers, and you know it! And what will we say to all the customers that ask about you? I think there's actually a lot of people who come here just because they know and like you!"

"Yes, I'll miss them for sure," said Lena, "but maybe some of them will come see me at the pancake house."

Assume you were Lena's supervisor. Do you think your customers will be affected by Lena's departure? How will you be affected personally? Do you think that Lena left King's for a "better" job? Explain your answer

Recruiting Team Members

After foodservice operators have thoughtfully designed their jobs, they will inevitably find they face the task of **recruiting** new team members. This may be the result of the opening of a new operation, staff expansion, or staff replacement.

Numerous factors affect recruiting activities in the foodservice industry. Examples include:

Key Term

Recruiting: The process of actively seeking out, finding, and hiring candidates for a specific position or job.

✓ *Legal constraints*—As addressed in Chapter 2, local, state, and federal laws significantly affect recruiting activities. Potential employers cannot seek out individuals based on non-job-related factors including age, gender, or physical attractiveness (if they do, significant legal problems are likely to result).

For example, some foodservice operators have historically viewed some positions as being best suited for men or women. Today, people of all genders can and do wait on tables, serve as bartenders, and complete back-of-house food production tasks.

Historically, the foodservice industry has provided tremendous opportunities for employees of all backgrounds. It will continue to do so, not only because it is the legal thing to do but also because it is the right thing to do.

✓ *Economic constraints*—Economic constraints can affect both the foodservice operator and job **applicants**. The pay rates offered to potential employees is typically determined by an operation's profitability. All foodservice operators face pay-related constraints that must be considered when recruiting.

Key Term

Applicants: Job searchers who have officially "applied" for a job opening.

Most applicants are attracted to or deterred from positions in part, at least, because of the compensation offered. The best interests of the employer and employee are served when pay rates and pay ranges for positions are reasonable and competitive. However, it is important to remember that pay rates are only one of several critical factors that good employees consider when applying for positions.

✓ *Industry perception constraints*—Some people incorrectly view the foodservice industry as one with few advancement opportunities and low-compensation rates. In fact, significant personal and financial rewards can be earned by employees with a variety of education and experience backgrounds. While a single operator is unlikely to change every applicant's perceptions, recruitment efforts should, when necessary, directly address potential candidate biases.

One way to do this is to focus on a position's varied and positive characteristics. These can include employment stability, work variety, an opportunity to use personal creativity in a team environment, the rewards of serving others, pleasant working conditions and, in some cases, the potential for promotions.

Each position in a foodservice operation has its own positive features and specific industry segments have their own unique and positive attributes. The attractive features of available positions should be publicized during recruiting because this can educate those who do not fully understand the advantages of a job or career in foodservice.

✓ *Organizational constraints*—Some applicants may react due to the type of foodservice operation advertising a vacant position. For example, operations in school foodservice units may be favorably perceived because these jobs have traditional hours and may offer above-average benefits including time off during summer months. Alternatively, those operating high-energy nightclubs will likely find their best potential applicants are drawn to the excitement of the operation despite working during nontraditional (late night) hours.

Foodservice operators should truthfully identify the employment advantages offered by their organizations. The reason: Those with excellent reputations as employers will generally have larger pools of potential employees than their counterparts with less favored reputations.

✓ *Position constraints*—Some foodservice jobs are perceived as highly desirable, and others are not and recruiting a large and qualified pool of applicants for the latter may be challenging. Many operators have difficulty filling manual labor positions such as dishwashers, janitors, and delivery drivers.

In job markets with a low-**unemployment rate**, a worker shortage may exist. Then, qualified applicants may be hard to find, and operators must work diligently and creatively to identify potential applicants.

Key Term

Unemployment rate: The number of unemployed people in a community or other designated area who are seeking work; expressed as a percentage of the area's entire labor force.

Despite the constraints to be considered, effective team member recruiting for foodservice operators can be successful when those with recruiting responsibilities focus on key applicant recruitment-related areas:

✓ Internal Recruiting
✓ Traditional External Recruitment Methods
✓ Internet-based External Recruitment Methods
✓ Recruitment of Diverse Team Members

Internal Recruiting

For many foodservice operators, candidate recruitment is an ongoing activity. One good way to address the team member search process is to consider it as either an internal or external process. An **internal search** is used when the best candidates for vacant positions are believed to be currently employed by the operation.

Key Term

Internal search: A promote-from-within recruitment approach used to identify qualified job applicants.

Done properly, an internal search approach can be very effective. For example, a foodservice operator needs a dining room supervisor and thinks the best job candidates are among the operation's current servers. A second example: A qualified lead cook can be selected from a pool of existing line cooks.

Current employees may be informed about pending job openings in individual conversations with potential applicants, in employee meetings, and by posting information on employee bulletin boards, Websites, newsletters, or other media. Advantages of using internal searches include that they

✓ build employee morale;
✓ can be initiated quickly;
✓ improve the probability of making a good selection because much is already known about the individual who will be selected;
✓ are less costly and time consuming than initiating external searches;
✓ result in reduced training time and fewer training costs because the individual selected need not learn about organizational topics with which they are already familiar;
✓ encourage talented individuals to stay with the organization.

Possible disadvantages to the use of internal searches could include:

✓ *Employee resentment*—When several employees have been considered for internal advancement and only one is chosen, those who were not chosen may feel resentment, and decreased morale can result.

✓ *Recruitment and selection are still needed*—When one position is filled internally, the position vacated must also be filled. Sometimes it may even be more difficult to attract qualified candidates for the newly created vacancy than to select an outsider for the original vacancy.

✓ *Increased use of training resources*—When two positions are filled with new employees (the original position and the now-vacant position), training efforts to address both employees' needs may be greater than if an external applicant is chosen for the original vacant position.

Despite these potential disadvantages, most foodservice operators believe the advantages of an internal search outweigh external searches. Most foodservice employees want to do a good job and improve themselves and their families through hard work and employer loyalty. Employers can reward them in a visible manner by considering, when possible, current workers for higher-level job openings.

While not technically an internal search system, some operators successfully utilize **employee referral** systems to discover potential applicants recommended by current employees.

Employee referral systems often work well because employees rarely recommend someone unless they feel the persons recommended can do a good job and fit well in the organization.

Key Term

Employee referral: A promote-from-within recruitment approach used to identify qualified job applicants.

Also, existing employees tend to have an accurate view of the job including the organization's culture, and this information reduces unrealistic expectations and can also help reduce new employee turnover. In some operations, a financial bonus is paid to a team member who recommends a person who is hired by and remains with the organization for a specified time.

There are potential challenges with employee referral systems including that those employees who recommend one or more persons may suggest friends or relatives regardless of their qualifications. For this reason, the same standards of employment consideration that apply to other individuals considered for a position should be used for all referred candidates.

Traditional External Recruitment Methods

Even when internal recruiting methods are effectively used, in many cases foodservice operators must conduct an **external search** outside their own organizations for new job applicants.

Key Term

External search: An approach to seeking job applicants that focuses on candidates not currently employed by the organization.

Commonly used strategies for external searches include:

✓ *Traditional advertisements*—These can range from simple "Help Wanted" or "Now Hiring" postings on-site or on public bulletin boards. Some venues for placing ads are free or low cost. These often include social media sites and bulletin boards at apartment complexes, childcare centers, supermarkets, libraries, community centers, and school newspapers. Employers seeking candidates for whom English is not their primary language may also target foreign-language newspapers and newsletters.

✓ *Educational institutions*—Most educational institutions provide services to help their graduates find jobs. Whether the new job requires a high-school diploma, specific vocational training, or a degree, educational institutions are a source of qualified job candidates that should not be overlooked. Technical schools and secondary (high) schools in an employer's area typically offer employers the chance, at little or no cost, to assess the quality of their students.

✓ *Unsolicited applications*—Foodservice operators often receive unsolicited applications or requests to be considered for job openings. These may arrive in an operator's office by letter, text, or e-mail, or by in-person delivery. Even if there are no current vacancies, these applications should be kept on file.

It is important to recognize, however, that unsolicited applications submitted to a foodservice operator have a relatively short life span because those who are searching for jobs typically continue their search until it ends successfully. Therefore, many operators review all unsolicited applications submitted each day, and some even arrange to conduct on-the-spot interviews when good candidates are scarce.

Find Out More

Some good recruiting news for foodservice operators is that many young people want to work in the restaurant industry. As a result, some of these young people enroll in vocational foodservice programs offered at the high school or community-college level.

In most of these programs, students add to their knowledge of safety and food sanitation as they deepen their understanding of food cooking methods and principles. Students in these programs gain knowledge that will help prepare them for jobs and careers in the restaurant and retail foodservice industries.

(Continued)

For foodservice operators, these programs provide opportunities to become involved with training at the local level. In some cases, internships can be arranged, or the operator can make class presentations.

Instructors in vocational foodservice programs often look to foodservice operators to keep them up-to-date on current trends in the industry. These instructors can be good sources of information about students in their programs who might fit the needs of a specific foodservice operator.

To learn more about these government funded training programs in your area, enter "Vocational foodservice training programs in (enter city or area)" in your favorite search engine and review the results.

Internet-based External Recruitment Methods

The Internet has significantly changed how foodservice operators recruit employees. Increasingly, popular social media sites are used to communicate with applicants about vacancies, questions, and other information. Also, many foodservice operators feature general "Career Opportunity" or "Current Openings" information on their websites along with more specific information about current position vacancies.

Career-related information on well-developed company websites may include:

✓ Organization overview, including press releases, current news stories, and awards
✓ Corporate culture including information about the organization's values, vision, and mission
✓ General information about the organization's history, operating statistics, and employee testimonials
✓ Current position vacancies
✓ Compensation ranges including a description of benefits

Websites may also include application forms and e-mail or other addresses for additional information. This is particularly useful for attracting younger workers who prefer online applications.

Many hospitality organizations utilize social media sites such as Uloop, Instagram, ZipRecruiter, LinkedIn, and Monster, and their own Facebook pages to offer information about employment opportunities. Twitter currently is another popular site that allows foodservice operators to announce position vacancies and receive inquiries from those who are potentially interested in the positions. For management-level openings, Hcareers is a site that provides a huge database of hospitality industry–related employment opportunities.

Technology at Work

Social media sites that emphasize and feature job openings are increasingly popular and especially so with younger workers. The popularity of these sites, however, can change very rapidly. As a result, foodservice operators must make it a point to stay up to date with changes in those social media sites that reflect increasing or decreasing popularity. Monitoring changes in such sites certainly need not be a daily activity, but it likely should be at least a monthly or quarterly activity.

There is little doubt that the popularity of specific social media sites and apps will continue to increase and decrease just as they have in the past. To learn more about the most currently popular social media sites used by job seekers, enter "Popular social media employment sites" in your favorite search engine and view the results.

Recruitment of Diverse Team Members

To compete effectively in recruiting market foodservice operators today, it is important to seek out diverse workers. Most foodservice operators recognize that **diversity** at work is a good thing, and they understand the importance of recruiting a diverse workforce. Race is often the first issue that comes to mind when diversity is considered, but there are many other factors that separate people into groups.

Key Term

Diversity: The term used to describe the variety of differences between people in an organization that includes race, gender, age, religion, sexual orientation, and more.

For foodservice operators, diversity should be viewed simply as differences among potential team members. These team member differences can come in many forms including:

✓ Gender
✓ Age
✓ Sexual orientation
✓ Race
✓ Ethnicity
✓ Weight
✓ Height
✓ Tattoos and body piercings
✓ Religious belief
✓ Educational level

✓ Socioeconomic status
✓ Physical conditions including;
  • Challenges related to general health
  • Challenges related to hearing
  • Challenges related to speech
  • Challenges related to vision
✓ Movement limitations (for example, the need for canes, walkers, and wheelchairs)
✓ Visible physical differences (for example, missing fingers, hands, arms, or legs)
✓ Differences in mental capacity

Some worker differences are the result of an individual's personal choices, and others are not. Either way, it is important to recognize that the presence of differences alone really do not tell much about the real character of a worker. In most cases, there is very little connection between visible differences that create diversity and those true differences that exist in each person's personality.

Experienced foodservice operators know that strength of character, a positive attitude, loyalty, and compassion for others can exist in diverse workers of any type. The real differences between good workers and less valuable workers are not based on what they look like, but on who they really are.

In most cases, visual differences demonstrate a person's individualism or show the specific physical challenges they face every day. Rarely do they tell much about a worker's true character. Personal reactions to seeing diversity do reveal something else, however. The way in which a person reacts to the visible difference of others often illuminates the true nature of their own character. For that reason, it is important to recognize that team members will often look to the actions of their own supervisors and operators as they decide for themselves how they will respond to workers of different abilities and appearances.

Devising and implementing strategies to recruit for a diverse workforce benefits foodservice operators directly because they increase the pool of viable candidates and contribute to a positive work environment.

Technology at Work

Recruiting software is an umbrella term for many different team member recruitment tools. It includes software used for advertising position vacancies, screening applications, and assessing an applicant's potential. They also help job candidates by simplifying the application process and facilitating clear communication with the organization that is hiring.

Foodservice operators use recruiting software to automate some parts of the recruitment process and to help secure sensitive applicant information that may be on job applications. How an operation appears online, the quality of the application process, communication preferences (some candidates prefer text messages instead of e-mails), and more come into play when attracting top talent. *Note:* This is especially so for younger generations who expect a compelling and mobile-friendly online application experience.

To review the features of some popular recruiting software programs, enter "Recruiting software for restaurants" in your favorite search engine and view the results.

Retaining Team Members

Experienced operators know that effective employee retention is just as important as effective recruiting. In fact, unless an operation emphasizes team member retention it will likely be in a constant state of endless recruiting.

In some cases, team members terminate their employment for reasons beyond the control of a foodservice operator. Examples include college students going back to school, employees who reach the retirement age, and those who move away to different cities or states. While factors such as these are beyond the control of an operator, creating a work environment that makes it easy for employees to want to remain part of the team is directly under the control of operators.

Experienced foodservice operators recognize that every team member wants to know their efforts at work have meaning and are valued. Knowing that this is so helps affirm an individual's self-worth and respect. As well, everyone wants to feel truly appreciated by those who benefit from their efforts and to know that this appreciation is put into words and actions. When they do, their response is overwhelmingly positive.

There are a number of specific activities foodservice operators can utilize to help create the positive work environment needed to encourage current workers to stay. These include:

1) Scheduling a 15-minute team meeting with hourly paid workers every month to share an operation's successes and goals to help team members stay connected to the operation's mission. These may be done by staff meetings, a morning doughnut and coffee get together, an afternoon cookie break, or other creative ways to strategically connect with, motivate, and praise team members.

2) Holding a private, closed-door meeting with each hourly paid employee every three months to clearly explain the importance of the team member's individual efforts to please guests and contribute to the success of the operation.

3) Meeting twice per year in a private, closed-door session with each hourly employee to personally thank him or her for a specific and positive aspect of their work performance. Examples could include good attendance, punctuality, consistently adhering to dress code requirements, and/or a positive attitude toward guests or fellow team members. Operators should find a positive trait in each team member and make appreciation for that trait known to them.

4) Annually sending a holiday card to the home address of each a team member with a personal (handwritten) note affirming the team member's importance to the success of the operation. *Note*: Recognize that these cards will be proudly displayed at the homes of many hourly paid workers.

5) Regularly planning small celebrations at work to recognize significant employee milestones such as one- or five-year anniversary, promotion, and similar accomplishments. During the celebration, emphasize to all workers the importance of loyalty and longevity and demonstrate appreciation with a modestly priced pin, gift, or other appropriate workplace reward.

6) Planning and implementing an annual team building exercise for after-work hours such as attending a sporting event or local festival, a hosted meal, or an afternoon trip to a coffee shop. During the event, emphasize the importance of teamwork to the team's success and express appreciation for each team member's contribution.

While each of these types of activities requires some of an operator's time, the retention benefits of the activities will generally far outweigh the time invested.

Recruiting and retaining a productive workforce is a key element in the successful operation of a foodservice business. Initially, interviewing and selecting the best applicants for vacant positions is also important, and doing so legally and effectively is the sole topic of the next chapter.

What Would You Do? 5.2

"I'm not stupid," said Kalina McCullough angrily to Tony Yabbara.

Tony was the owner/operator of Fun House Pizza, a combination pizza parlor and video arcade that catered to families and was a popular spot for hosting children's birthday parties, sports team celebrations, and similar events.

Kalina worked the front counter of Fun House Pizza as an order taker and cashier. Fun House was very popular and almost always busy. Today, Tony couldn't tell if Kalina was about to cry or kick the door to the office where they were meeting. Either way he could tell she was very upset.

The problem had started when a guest Kalina had been working with demanded to see the manager. The customer's debit card had been rejected by the operation's point-of-sale payment card integration system. As Kalina tried to explain that, the guest became increasingly agitated and then loudly demanded to see the manager.

When Tony arrived at the cashier's stand to address the guest's issue, he asked what he could do to assist. The guest immediately replied "Well, you might start by firing the employees you have who are too stupid to process a simple credit card transaction!"

It was at that point that Tony had asked Kalina to go to the office and wait for him. After finishing with the guest, Tony had entered the office to find a still visibly upset Kalina.

"I know that," replied Tony to Kalina's opening comment. "I also know that there are a lot of business people who like to say the customer is always right! Well, the customer is *not* always right, and we need to talk about this one."

Assume you were Tony. Do you think the support you show for Kalina in this meeting will be important to how she views you and her job? Do you think that it is likely that whatever you say to Kalina in this meeting will be shared by her with her fellow co-workers? How could that affect how other workers view the work environment at Fun House Pizza?

Key Terms

Task list	Open-door	Unemployment rate
Position analysis	communication	Internal search
Performance standards	(policy)	Employee referral
Tip credit	Recruiting	External search
Pay range	Applicants	Diversity

Operator's 10-Point Tactics for Success Checklist

Evaluate your need for, and the current status of, each of the following operational tactics. For those tactics you think are important, but not yet in place, develop an action plan for its implementation including who will be responsible for the tactic's completion and the target date by which it should be completed.

Tactic	Don't Agree (Not Done)	Agree (Done)	Agree (Not Done)	If Not Done	
				Who Is Responsible?	Target Completion Date
1) Operator recognizes that it is very important to successfully build a team of qualified employees.	———	———	———		
2) Operator understands the relationship between effective job design and the ability to attract qualified job applicants.	———	———	———		
3) Operator recognizes the legal, economic, industry perception of, organizational, and positional restraints associated with effective employee recruitment.	———	———	———		
4) Operator can state the potential advantages that can accrue from the use of internal recruiting.	———	———	———		
5) Operator can state the potential disadvantages that can accrue from the use of internal recruiting.	———	———	———		
6) Operator can list specific targeting strategies best used in a traditional external search for job applicants.	———	———	———		

Tactic	Don't Agree (Not Done)	Agree (Done)	Agree (Not Done)	If Not Done	
				Who Is Responsible?	Target Completion Date
7) Operator understands the increasing use of the internet as an effective vehicle for conducting searches for qualified job applicants.	———	———	———		
8) Operator understands the importance of recruiting a diverse workforce when seeking to expand the pool of qualified applicants.	———	———	———		
9) Operator recognizes that the real differences between good workers and less valuable workers are not based on what they look like, but rather on their character.	———	———	———		
10) Operator has considered the use of specific weekly, quarterly, and annual activities that can be utilized to enhance employee retention.	———	———	———		

6

Interviewing and Selecting Team Members

What You Will Learn

1) Tools Used to Help Select Qualified Team Members
2) The Dangers of Negligent Hiring
3) How to Make a Job Offer
4) Documentation Required for New Team Members

Operator's Brief

Some of a foodservice operator's most important decisions relate to the selection of new team members. In this chapter, you will learn how to choose new team members in a manner that is legally sound and designed to help ensure only high-quality team members are selected.

Foodservice operators choosing new team members have a number of important selection-related tools available to them. These include formal job applications that should be completed by all candidates for employment. Employment interviews are another important selection tool, and in this chapter, you will learn about applications as well as interview questions that are and are not permitted to be asked when conducting employment interviews.

In some cases, you may utilize pre-employment testing in your employee selection process and in this chapter, you will learn how that can best be done. For some select positions, foodservice operators may also conduct background checks and seek references. Both of these important tools are examined in detail in this chapter.

While in most cases you will be free to hire any new team member you choose, negligent hiring, a failure to exercise reasonable care as new employees are selected and retained, is a key concept and, in this chapter, you will learn how best to avoid it.

A formal job offer letter is the beginning of a legal contract between an employer and a new employee. In this chapter, you will learn about the information that should be included in job offer letters to best protect yourself and your operation. Your final step in choosing a new employee is the completion of required documents. These documents include Federal forms I-9 and W-4, and in some cases, required state and local tax forms.

CHAPTER OUTLINE

Choosing Valuable Team Members
    Applications
    Interviews
    Pre-employment Testing
    Background Checks and References
Negligent Hiring
Making Job Offers
Required Documentation for New Team Members
    Form I-9 (Federal form)
    Form W-4 (Federal form)
    State and Local Tax Forms

Choosing Valuable Team Members

The proper selection of new team members is one of a foodservice operator's most important tasks. There are numerous advantages that accrue to operators when they select new employees properly. Effective hiring:

Saves time: When the right employee is hired the first time, it saves the time and effort operators must otherwise spend to recruit and interview. When newly employed workers have the proper characteristics, they can more quickly and effectively understand their new roles and responsibilities, and this means that less time is spent on training.

Decreases hiring costs: There are real costs to hiring new workers. Those costs may include advertising vacancies and various administrative costs of hiring. Choosing the right new employee the first time helps to minimize hiring costs.

Reduces burn out of existing employees: In most cases, foodservice operators hire new employees because they are immediately needed. When the right employee is hired, the new team member quickly contributes to the team. As this occurs, existing staff may have to work fewer overtime hours, and the more experienced staff can more easily handle excessive workloads while new staff members learn about tasks they will be required to do.

Enhances morale: When new employees of high quality are hired, they can quickly adjust and contribute their vital skills and experience. As a result, they encourage other employees in their own jobs. Alternatively, a bad hire can have the opposite effect. Poor hires, including those with negative attitudes and unsatisfactory work ethics, can detract from an operation's culture and potentially frustrate current employees.

After foodservice operators use their chosen techniques to recruit and identify a pool of qualified candidates (see Chapter 5), they must select the best applicants to fill their vacant positions. To do so they typically use some or all of the following employee selection tools:

Applications
Interviews
Pre-employment Testing
Background Checks and References

Applications

An employment **application** should be completed by all candidates for employment, and application forms which are most effective do not need to be extremely lengthy or complex.

The purpose of an application is to help foodservice operators learn information required to determine whether the applicant can, with appropriate training, perform a job's essential functions. The requirements for a legitimate, legally sound application are many but, in general, the questions asked should focus exclusively on job qualifications. It is always a good idea for foodservice operators to have their proposed employment applications reviewed by an attorney who specializes in employment law before they are put into use.

Each candidate for a specific position should fill out an identical application, and the applications should be kept on file. It is always a good practice for the application to clearly state the **at-will** nature of the employment relationship.

Key Term

Application: A form, questionnaire, or similar document that applicants for employment are required by an employer to complete. An application may exist in a hard copy, electronic copy, or Internet medium.

Key Term

At-will (employment): An employment relationship in which an employer can, at any time, terminate the relationship without liability as long as it is not discriminatory. Likewise, an employee is free to leave a job at any time for any or no reason with no adverse legal consequences.

In 49 U.S. states, at-will employment is permitted. The only exception is Montana, which generally requires employers to have **good cause** for dismissing an employee who has worked through a stated probationary period.

In most cases, a foodservice operator uses the same application for all job applicants. Doing so helps create and ensure a structured and consistent assessment process that most often begins with an application review, and then proceeds through one or more rounds of interviews.

Key Term

Good cause: A reason for taking an action or failing to take an action that is reasonable and justified when viewed in the context of all surrounding circumstances.

A job application may ask for general information about previous job experiences, special skills, and education. It may also ask specific questions about a particular job. For example, an operator may use a job application to ask an applicant about their number of years of work experience or whether they are adept at using a particular piece of kitchen equipment.

They may also ask about the applicant's willingness to work specific hours. Questions such as these can help an operator to narrow down a large pool of applicants to one in which each applicant meets minimum position requirements. The use of uniform applications also enables easy comparisons of one applicant to another in a fair and consistent manner. A well-constructed application should also have a signature section that allows applicants to sign the document indicating all information they have provided is true and accurate to the best of their knowledge.

Technology at Work

To compete effectively in an increasingly tight labor market, foodservice operators should make it easy for potential employees to apply for a job. One good way to do this is to ensure that an operation has an easy-to-use online job application program.

An on-line job application is a form on the internet where applicants can communicate their skills and relevant experience for a specific job or position. Foodservice operators use online applications to help speed up the hiring process and increase the number of potential job candidates for the position.

Several technology companies offer online application services appropriate for small businesses. To learn more about the ways online application programs work and can be publicized to potential job applicants, enter "Online job application programs for restaurants" in your favorite search engine and view the results.

Many applicants for jobs in the foodservice industry may have limited skills in speaking, reading, or writing English. These limitations should not necessarily disqualify an applicant from being selected and doing an excellent job. Knowing this, many foodservice operators provide assistance to applicants who need help when filling out their employment applications.

Most foodservice operators use a single, uniform application form for all job candidates to help show who is considered an "applicant." This is an important determination in complying with the federal government's record retention and reporting requirements. For example, Title VII of the Civil Rights Act requires covered employers to retain "applications" for employment and other documents pertaining to hiring for one year from the date the records were made, or the last action was taken. Does that mean a foodservice operator who receives unsolicited resumes from a job seeker must keep them for one year? What if dozens or even hundreds of such e-mails are received? Are they all really applicants?

The EEOC (see Chapter 2) and the Office of Federal Contract Compliance Programs broadly define *applicant* to include any person who has indicated an interest in being considered for hiring, promotion, or other employment opportunities. This interest might be expressed by completing an application form in writing or verbally, depending on the employer's practice.

Increasingly, foodservice operators receive job applications from those who have applied for a job online. Fortunately, the EEOC has issued opinions to clarify recordkeeping for applicants using the Internet and related cyber technologies. For example, EEOC's guidance limits the definition of *applicant* in the context of the Internet and related technologies. It does so by indicating persons who have indicated an interest in a specific position that the employer has acted to fill and who have followed the employer's standard procedures for submitting an application. For this reason, many employers require that *all* job applicants, including those with formal resumes and/or letter of inquiry, submit a completed job application to be considered for employment, and these completed applications are then retained.

Figure 6.1 is an example of an employment application that is legally sound (in most states). Note specifically how questions relate only to the ability to work, work history, and job qualifications.

Interviews

Some job candidates who have submitted employment applications will be selected for one or more interviews. These can be powerful selection tools; however, the types of questions that can legally be asked in an interview are restricted. If interviews are improperly performed, significant legal liability can result and, if an applicant is not hired based on their answer to, or refusal to answer, an

Lighthouse Restaurant

Application for Employment

Name: _____, _____ _____
           Last   First   Middle

Street Address _____, _____ _____
                   City   State   Zip

Telephone (_____) _____-_____ Social Security # _____-_____-_____

Position Applied for _____

How did you hear of this opening? _____

When can you start?: _____/_____/_____

Desired Hourly Wage: $_____

	Yes	No
• Are you a U.S. Citizen or otherwise authorized to work in the U.S.?	☐	☐
• Are you capable of performing the essential functions of the job you are applying for with or without reasonable accommodation?	☐	☐
• If applying for a job involving the service of alcoholic beverages, are you over the age of 21?	☐	☐
• Are you under the age of 18? Birth Date: _____/_____/_____	☐	☐
• Are you looking for full-time employment?:	☐	☐

- If no, what days and hours are you available? (please list all that apply)

	From	To
Monday		
Tuesday		
Wednesday		
Thursday		
Friday		
Saturday		
Sunday		

	Yes	No
• Do you have dependable means of transportation to and from work?	☐	☐
• Do you have any criminal charges pending against you?	☐	☐
• Have you been convicted of a felony in the past seven years?	☐	☐

If yes, please fully describe the charges and disposition of the case:

_____

_____

(Note: Conviction of a felony will not necessarily disqualify you from employment)

Education:

	School Name/ Location	Year Completed	Major/ Degree
High School:			
Technical School:			
College:			
Other:			

Figure 6.1 Sample Employment Application

In addition to your work history, are there other certifications, skills, qualification we should know about. _____
_____

Employment History. (start with the most recent employer)

Company Name: _____
Location: _____
Starting Position: _____
Start Date: _____/_____/_____   Starting Wage: $_____
Ending Position: _____
End Date: _____/_____/_____   Ending Wage: $_____
Name of Supervisor: _____
　　　　　　　　May we contact: Yes _____ No. _____
Responsibilities:
_____
_____

Reason for Leaving:
_____
_____

Company Name: _____
Location: _____
Starting Position: _____
Start Date: _____/_____/_____   Starting Wage: $_____
Ending Position: _____
End Date: _____/_____/_____   Ending Wage: $_____
Name of Supervisor: _____
　　　　　　　　May we contact: Yes _____ No. _____
Responsibilities:
_____
_____

Reason for Leaving:
_____
_____

I state that the facts written on this application are true and complete to the best of my knowledge. I understand that if I am employed, false statements on this application can be considered cause for dismissal. The company is hereby authorized to make any investigations of my prior educational and employment history. I understand that employment at this company is "at will" which means that I or the company can terminate the employment relationship at any time, with or without prior notice. I understand that no supervisor, manager, or executive of this company, other than its owner has the authority to alter the at will status of my employment.

I authorize you to make such legal investigations and inquiries into my personal and employment, criminal history, driving record, and other job-related matters as may be necessary in determining an employment decision.

_____   _____

Signature   Date

Confidential material/property of Lighthouse Restaurant, LLC

Figure 6.1 *(Continued)*

inappropriate question, that applicant may have the right to file (and maybe win!) a lawsuit.

The EEOC suggests an employer consider three issues when deciding whether to include a particular question on an employment application or in a job interview:

1) Does this question tend to screen out minorities or women?
2) Is the answer needed to judge this individual's competence for job performance?
3) Are there alternative, nondiscriminatory ways to judge the person's qualifications?

Foodservice operators, or those they designate to make hiring decisions, must carefully select questions to ask in an interview. In all cases, it is important to remember that the job itself dictates whether questions are allowable. The questions to be asked of all applicants should be written down in advance and carefully addressed. In addition, supervisors, co-workers, and others who may participate in the interview process should be trained to avoid asking inappropriate questions that could increase an operation's potential liability.

In general, age is considered irrelevant in most hiring decisions, and questions about date of birth are improper. However, age can be a sensitive pre-employment question because the Age Discrimination in Employment Act protects employees 40 years old and above. Asking applicants to state their age if they are younger than 18 years old is permissible because young persons are permitted to work only a limited number of hours each week. It may also be important when hiring bartenders and other servers of alcohol to confirm that their ages are at or above the state's minimum age for serving alcohol.

Questions about race, religion, and national origin are always inappropriate, as is the practice of requiring that photographs of the applicant be submitted before or after an interview. Questions about physical traits such as height and weight violate the law because they may eliminate a disproportionate number of female, Asian-American, and Spanish-surnamed applicants who are statistically shorter than White males.

If a job does not require a specific level of education, asking questions about educational background may be improper. Applicants can be asked about their education and credentials if these are bona fide occupational qualifications (see Chapter 2). For example, asking an applicant who is a candidate for an Executive Chef's position if they have graduated from an American Culinary Federation (ACF) recognized culinary school is allowable. Similarly, asking an applicant if they use illegal drugs or smokes is permissible because either of these traits can be legally used to disqualify applicants. Asking candidates if they are willing to submit to a voluntary drug test as a condition of employment is also allowable.

Federal, state, and local laws can impact how employees are selected. The Americans with Disabilities Act (ADA; see Chapter 2) provides a good example of how foodservice operators are directly affected by employment-related legislation.

The ADA-related actions to be taken begin even before an employee is hired because some special provisions may be necessary for employment interviewing. These can include:

Scheduling an on-site interview with a qualified applicant with a hearing loss rather than requiring the person to first pass a telephone screening interview.

Modifying the job application process so a person with a disability can apply. Examples include providing large print, audio tape, or Braille versions of the application or allowing a person to apply with a paper application when an online application is normally required (or the reverse).

Providing a sign language interpreter or a reader during the interview process.

Conducting interviews in a first-floor office when an elevator is unavailable and ensuring all areas required for the application process are accessible.

Altering the format or the time allotted for a required test unless it is measuring a skill that is essential to job function.

Providing or modifying equipment or tools needed to perform an essential function of the job when the function is tested or assessed as part of the job application process.

Medical exams as part of the employee selection process are also addressed by the ADA: They are prohibited before a job offer is made. After a job has been offered and before employment begins, a medical examination may be required, and the job offer may be a condition of the exam results. Note that an examination must be required of every applicant in the same job category.

If the employment offer is withdrawn because of medical findings, the employer must show the rejection was job-related because of a business necessity, and there was no reasonable accommodation that enables the individual to perform that job's essential functions.

The ADA does not generally allow employers to require medical examinations of employees except:

To determine whether the employee can do the essential job functions after a leave for illness or injury or if the employee's fitness for duty is questioned;

After an employee requests an accommodation to determine whether the employee has an ADA-related disability and what reasonable accommodations may be required;

If required for employer-provided health or life insurance or for voluntary participation in an employer-sponsored health program; or

If it is required by a federal law or regulation.

Passage of the ADA also changed questions that can be asked of a job applicant. For example, questions that should *not* be asked on job applications or during interviews because they likely violate the ADA include:

1) Have you ever been hospitalized?
2) Are you taking prescription drugs?
3) Have you ever been treated for drug addiction or alcoholism?
4) Have you ever filed a workers' compensation insurance claim?
5) Do you have any physical defects, disabilities, or impairments that may affect your performance in the position for which you are applying?

As they conduct their employment interviews, foodservice operators must recognize that safe questions can be asked about an applicant's present employment, former employment, and job references. Questions asked on the application and during the interview should focus on the applicant's job skills and nothing more.

Find Out More

Very few foodservice operators would intentionally ask a question of an applicant that is illegal. Sometimes, however, avoiding the asking of illegal questions can be tricky.

For example, assume an interview is taking place on a Sunday. The interviewer greets the applicant by saying "Welcome, how are you today?"

The applicant replies, "I'm doing great, I just got out of church!"

While it might seem reasonable for the interviewer to follow up with a "Good for you, What church do you attend?" query, and such a question might be considered normal and even friendly, it would also be illegal!

Asking an applicant about the church they attend is prohibited under Title VII of the Civil Rights Act because it could potentially make religion (and church attendance!) a factor in the interviewer's job offer decision making.

One good way to become more familiar with questions that are, and are not, appropriate to be asked in an interview is to watch videos on YouTube that address this key issue.

To learn about questions that can and cannot be asked of job applicants, go to the YouTube website, and enter "Illegal job interview questions to avoid asking" in the YouTube search bar, and then watch one or more of the posted videos addressing this important topic.

What Would You Do? 6.1

"I liked her personality a lot, but I just didn't like her looks," said Margie, the cafeteria supervisor.

"What didn't you like?" asked Terri, the head Dietitian for the hospital food-service where Margie worked and the hiring manager for the hospital's food-service operation.

Margie and Terri were discussing Aurora, a 20-something young lady who had just finished an interview for a job bussing tables in the hospital's "Open to the public" dining room.

"Well, I don't know if its 'Goth,' or 'Emo,' or whatever kids are calling it today, but all that black around her eyes and on her lips, and the multiple ear piercings. All those hoops! It just seems very odd to me," said Margie.

"I agree it looks different," said Teri, "but I remember when I was about her age my parents thought the way I dressed was outlandish! Now I would guess they would call the outfits I wore back then old school!"

"Well, she wouldn't be in violation of our no visible tattoo policy," said Margie, "because she didn't have any facial tattoos. But maybe we need a better policy on appropriate dress?"

"Well, we certainly have the legal right to establish an appropriate dress code," said Terri, "but if we get into clothing colors, earrings, eye make-up styles, and lipstick colors, I think we might start to be on a slippery slope with many of our other employees!"

Assume you were Terri. Employers are free to regulate the physical appearance of their employees as long as it is done in a non-discriminatory and equitable way, and they are free to refuse hiring employees that do not conform to standards addressed in company policy. Assuming she was qualified, would you be in favor of hiring Aurora? Do you agree with Margie that you need a policy on appropriate dress that specifically addresses clothing and make-up colors? Explain your answer.

Pre-employment Testing

Pre-employment testing can improve the employee screening process because, for example, test results can measure the relative strengths of two applicants. In the foodservice industry, pre-employment testing is generally of three types: skills, psychological, and drug screening.

Skills tests can include activities such as typing tests for office workers, computer application tests for those involved in managing websites, and, for chefs and cooks, food production tasks.

Psychological testing can include personality tests and others designed to predict performance or mental ability. For skill and psychological tests, it is critical to remember that, if the test does not have documented **validity** and **reliability**, its results should not be used to affect hiring decisions.

Pre-employment drug testing is allowable in most states, and this can be an effective tool for reducing insurance rates and potential worker liability issues. Many foodservice operators believe a drug-free environment attracts better applicants with the resulting effect of a higher-quality workforce. If pre-employment drug testing is used, care is needed to ensure the accuracy of results. In some cases, applicants whose erroneous test results have cost them a job have successfully sued the employer.

Key Term

Validity (test): The ability of a test to evaluate only what it is supposed to evaluate.

Key Term

Reliability (test): The ability of a test to yield consistent results.

The laws surrounding mandatory drug testing are complex. Foodservice operators who elect to implement a voluntary or mandatory pre- or postemployment drug testing program should first seek advice from an attorney who specializes in labor employment law in the jurisdiction in which the operator does business.

Background Checks and References

Foodservice operators sometimes use background checks before hiring workers for selected positions because many resumes and employment applications are, at least partially, falsified. While many types of background checks are available, not all are advisable. Background checks should be specifically tailored to obtain only information relating directly to each applicant's employment suitability.

Commonly used background checks include:

Criminal history—As a general rule, criminal conviction records should be checked when there is a possibility that the person could create significant safety or security risks for co-workers, guests, or others. Examples include employees who will have close contact with minors, the elderly, the disabled, or patients, and those who will have access to weapons, drugs, chemicals, or other potentially dangerous materials.

Other examples include applicants who will work in or deliver goods to customers' homes, and those who will handle money or other valuables or have access to financial information or employees' personal information. In addition, some states require a check for criminal convictions before hiring individuals as employees of healthcare facilities (including foodservices), financial institutions, or public schools.

Credit reports—Credit reports typically include financial information such as payment history, delinquencies, amounts owed, liens, and judgments relating to an applicant's credit standing. Arbitrary reliance on the results of these checks, however, has sometimes been found to result in adverse impact discrimination against women and minorities.

Accordingly, use of credit reports should be limited to situations where there is a legitimate business justification. Examples can include jobs that entail monetary responsibilities, the use of financial discretion, or similar security risks.

Driving records—Motor vehicle records (MVRs) are available from state motor vehicle departments. They usually contain information about traffic violations, license status, and expiration dates. MVRs should be checked for any employee who will drive a company or personal vehicle for the employer's business including making food deliveries to guests.

Academic credentials and licenses—While less often addressed, academic information such as schools attended, degrees awarded, and transcripts should be verified when a specified level or type of education is necessary for a particular job. Similarly, proof of licenses and their current status, expiration dates, and any past or pending disciplinary actions should be obtained if a license is required for the position in question.

If background checks are to be used, four key principles should always be addressed to ensure they are completed legally and effectively.

Principle 1: Always obtain written consent before conducting any background check. This helps protect against invasion of privacy, defamation, and other wrongful act claims. It is also a good idea to expand the waiver language on consent forms to include the employer and those who assist with background checks including staff, former employers, and screening firms.

Principle 2: Evaluate the results fairly and consistently. Operators must avoid hasty applicant rejections when negative information surfaces during a background check. They must consider the negative information in the context of the job to be performed. For example, to reject an otherwise qualified cook because of a poor driving record is unwise because the job requires no business driving. However, it may be a sensible action for a food delivery driver.

Principle 3: Restrict access to information obtained in background checks. This information should be kept in secure, confidential files and disclosed only on a strict "need to know" basis. Access to records relating to criminal or financial history should be limited as narrowly as possible.

Principle 4: It is wise to do background checks as one of the last steps in the selection process. The reason: there is no need to spend the time and money for background checks on an applicant until the decision has been made to offer the applicant a position.

Using background checks as a screening device does involve some risk to and responsibility for employers. Seeking only information with a direct bearing on the position for which an applicant is applying is important. If an applicant is denied employment based on a background check, the employer should give the applicant a copy of that report. Sometimes applicants can help verify or explain the results of background checks. Also, reporting agencies can make mistakes and, if false information influences a hiring decision, the foodservice operation may be put at legal risk.

In addition to background checks, some foodservice operators rely on references when making decisions. However, seeking an applicant's **references** (and reliance upon them), is relatively controversial in today's workplace.

Key Term

Reference (employment): The positive or negative comments about an employee's job performance provided to a prospective employer.

Employment references have historically been a popular applicant screening tool. However, in today's society where lawsuits are often frequent, information of this type is often more difficult to obtain. One reason is because employers may be held liable for inaccurate comments made about past employees. This concern is increased because job seekers can employ the services of companies that specialize in providing job seekers with a confidential and comprehensive verification of the employment references given by former employers. Therefore, the information provided in a reference is now generally limited to the employer's name, employee's name, date(s) of employment, job title, and the name and title of the person supplying the information.

If references from past employers are sought, it is best to secure the applicant's permission in writing before contacting a former employer to help minimize the risk of litigation related to the reference checks.

Foodservice operators themselves must also be cautious in both giving and receiving reference information. Employers are usually protected if they give a truthful reference; however, they may still incur the time and expense to defend against charges of unfairness brought by a former employee. Consider an employer who provides a reference stating a former employee was terminated because they "didn't get along" with co-workers. The employer may need to prove that statement is true and, at the least, the former employee created all the difficulties.

To minimize the chances of a successful lawsuit, foodservice operators should never reply to a request for information about a former employee without a copy of that employee's signed release authorizing the reference check. The amount of information disclosed varies by individual case, but answers should be honest and defendable. It is also best to not disclose personal information such as marital or financial problems because this could result in an invasion of privacy lawsuit.

If an applicant provides letters of reference, it is often a good idea to contact the authors of the reference letters to ensure that they did write them before the references are used to impact a hiring decision.

Negligent Hiring

Providing references for past employees could subject foodservice operators to litigation if the comments made are challenged, if the information secured is false, and/or if it is improperly disclosed to third parties and violates the employee's right to privacy. However, failure to conduct background checks on applicants for some positions can subject operators to even more legal difficulty under the doctrines of **negligent hiring** and **negligent retention**.

> **Key Term**
>
> **Negligent hiring:** Failure of an employer to exercise reasonable care in the selection of employees.

> **Key Term**
>
> **Negligent retention:** Retaining an employee after the employer becomes aware of an employee's unsuitability for a job by failing to act on that knowledge.

Negligent hiring liability usually occurs when an employee who caused injury or harm had a reputation or record that showed their propensity to do so, and this record would have been easily discoverable if reasonable care (a diligent search) had been shown. Similarly, negligent retention may be charged if an employer hires an employee and then discovers disqualifying information but does not remove the employee from the job.

How can foodservice operators show reasonable care was used when hiring? The best tactic is to verify all pertinent information about each candidate before making a job offer. For example, it would be difficult to argue that the employer knew or should have known about false information if multiple references were contacted and if all relevant applicant references were verified to the best of an employer's ability. It would also be difficult for a judge or jury to discover an employer had been negligent in its hiring processes if this standard of care were exercised before every hiring decision.

Making Job Offers

A final step in the selection process is to clarify the conditions of the **employment agreement** with the new employee.

All employers and employees have employment agreements with each other. They can be as

> **Key Term**
>
> **Employment agreement:** The terms of the employment relationship between an employer and employee that specifies the rights and obligations of each party.

simple as agreeing to a specific hourly wage rate and at-will employment for both parties. This can be true even if nothing is in writing or if work conditions were not discussed in detail.

Employment agreements may be individual and cover only one employee, or, as in a unionized operation, they may involve groups of employees. In general, hospitality industry employment agreements are established orally or with a formal **job offer letter**.

Properly composed job offer letters can help prevent legal difficulties caused by employee and/or employer misunderstandings because they detail specific terms of the offer made by the employer to the employee.

Key Term

Job offer letter: A proposal by an employer to a potential employee that specifies employment terms. A legally valid acceptance of the offer creates a binding employment agreement.

Some foodservice operators believe job offer letters should only be used for management positions but, to avoid difficulties, all employees should have signed job offer letters in their personal files. Components of a sound job offer letter include:

- ✓ Position being offered
- ✓ Compensation, including benefits
- ✓ Evaluation period and compensation review schedule
- ✓ Start date
- ✓ Location of employment
- ✓ Special conditions such as the at-will relationship
- ✓ Reference to the employee handbook (see Chapter 5) as a source of information about employer policies governing the workplace
- ✓ Lines for the employer and employee signatures
- ✓ Date of the signatures

A job offer letter may be conditional or final. With a conditional offer letter, the employer tentatively (conditionally) offers the job subject to "conditions" that must be met before the job offer is finalized (examples include passing drug tests and/or background checks).

When a conditional job offer letter is used, the legally binding employment agreement is not in effect until the employee accepts the terms of the job offer letter and fulfills the requirements it identifies. In contrast, a final job offer letter contains no conditions to be met before acceptance. An enforceable employment contract is in effect when the final job offer letter is legally accepted by the job applicant.

To illustrate the difference between a conditional and final job offer, consider Antonio, who applies for a job as a full-time bartender. He is selected and given a conditional job offer letter that specifies the employment condition that he must pass a drug test. While Antonio can sign the letter when it is received, his employment will not be finalized until he passes the drug test.

Required Documentation for New Team Members

Even after an employment offer has been made and accepted, foodservice operators have one final task to complete before an applicant can become an "official" team member. That task involves obtaining the proper documentation for the new worker.

Required documentation for newly chosen team members include:

✓ Form I-9 (Federal form)
✓ Form W-4 (Federal form)
✓ State and Local Tax Forms

Form I-9 (Federal Form)

All newly hired employees are required to fill out an Employment Eligibility Verification form (commonly known as an I-9 form) stating that they are authorized to work in the United States. U.S. Citizenship and Immigration Service (USCIS) regulations allow an individual 72 hours from time of hire in which to complete Form I-9.

The form requires potential employees to verify both their citizenship status and legal eligibility to work. There are numerous documents that an employee can use to provide this information including a U.S. passport, driver's license, Social Security card, and school identification card.

Under current law, employers are not required to verify the authenticity of the identification documents they were presented, but they must keep a copy of them on file. The documents must pass a good-faith test ("Do they look real?") and, if they do, the applicant may be hired.

If an employee is later found to be unauthorized, the employer must terminate the worker's employment. Employers who do not end employment of unauthorized workers or who knowingly hire unauthorized workers can face significant fines.

Find Out More

E-Verify, authorized by the Illegal Immigration Reform and Immigrant Responsibility Act of 1996 (IIRIRA), is a web-based system through which employers electronically confirm the employment eligibility of their new employees.

In the E-Verify process, employers create "cases" (individual employee records) based on information taken from an employee's Form I-9, Employment Eligibility Verification.

E-Verify then electronically compares that information to records available to the U.S. Department of Homeland Security (DHS) and the Social Security Administration (SSA). The employer usually receives a response within a few seconds that either confirms the employee's employment eligibility or indicates the employee needs to take further action to complete the case.

E-Verify is administered by Social Security Administration and U.S. Citizenship and Immigration Services (USCIS). USCIS facilitates compliance with U.S. immigration law by providing E-Verify program support, user support, training and outreach, and developing innovative technological solutions in employment eligibility verification.

The use of E-Verify is not mandatory (except in certain states), but its use is always a good idea.

To find out more about the E-Verify program and how it helps employers stay in compliance with the law, enter "Why use E-Verify?" in your favorite search engine and review the results.

Form W-4 (Federal Form)

When a foodservice operator hires a new employee, they must also ensure the employee completes a Federal Form W-4. This form (formally titled "Employee's Withholding Certificate") is an IRS form that employees fill out and submit to their employers when they begin a new job.

Essentially, employers use the information provided on a W-4 to calculate how much tax to withhold from an employee's paycheck throughout the year. Form W-4 tells a foodservice operator (the employer), about the employee's filing status, multiple jobs adjustments, amount of credits, amount of other income, amount of deductions, and any additional amount to withhold from each paycheck. These are important because the data is used to compute the amount of federal income tax to deduct and withhold from the employee's pay. If an employee fails to submit a properly completed Form W-4, the employer must withhold federal income taxes from their wages as if they were single or married filing separately.

Technology at Work

Foodservice owners, operators, and managers who must complete weekly or bi-weekly payroll tasks typically use payroll software, an online payroll service, or a spreadsheet to calculate the amounts to be paid and the withholding and taxes to be withheld from their workers' pay.

This task can become more complex when a foodservice operation has many employees, or when turnover is very high. It also becomes more complex when large numbers of tipped workers are employed because operators are required to withhold taxes on tips.

Because of its complexity, increasing numbers of foodservice operators elect to choose a payroll service or purchase payroll software designed specifically for restaurants.

Fortunately, numerous companies create software specifically for preparing payroll in foodservice operations. To review the product and service offerings of some of these, enter "Most popular payroll programs for restaurants" in your favorite search engine and view the results.

State and Local Tax Forms

In addition to federal income taxes, workers in many states must pay various state (and sometimes local) income taxes as well. Currently, eight states in the United States choose not to impose an income tax.

In the majority of states, however, income taxes are assessed in a manner similar to the federal government's approach. But slight differences do exist in some states and, when they do, the state may require employers to utilize a separate form that is typically like the Federal Form W-4. If the state in which a foodservice operation conducts business calculates personal income tax using a **flat rate**, rather than a **graduated rate**, the required tax form will not closely resemble the federal form.

Key Term

Flat rate (income tax):
A taxing system in which each taxpayer pays the same percentage of their income in taxes.

Key Term

Graduated rate (income tax):
A taxing system in which higher tax rates are applied to higher income workers.

After new team members have been selected and all required documentation is completed, the new team member is ready to begin work. The first few days (and even the first few hours!) of employment can be critical to the long-term success and retention of new employees.

As result, how foodservice operators structure the initial days of work for new team members including what the new team members will learn and do is critical and will be the sole topic of the next chapter.

What Would You Do? 6.2

"He's super qualified for the job, and his interview went great, I don't see the problem," said Frank Ginzer, the assistant manager at Fayyad Brothers Po-Boy shop, located in a busy strip mall shopping center a half mile from the State's capitol building.

Frank was talking to Sandra Basset, the shop's general manager. Frank and Sandra were discussing Stan Burt, an applicant for a full-time job working in the shop's production kitchen.

Both Frank and Sandra had interviewed Stan. His work history was good, and he certainly had the skills needed to do the job they had for him.

"Well, my problem," said Sandra, "are his Facebook and X accounts. I went on both of them because he doesn't have them set to private, and what I saw was pretty concerning."

"What did you see?" asked Frank.

"Well, quite a lot really," replied Sandra, "There were a lot of recent postings, and many of them were typed in capital letters. Mostly they were discussing how bad the direction of the country was, how secret groups and sinister forces were plotting to overthrow America, and the need to take-up arms, if necessary, to prevent the takeover. I'm not sure I really understood all of it, but I thought it was a little scary to tell you the truth."

"Wow!," said Frank, "that really is interesting!"

Assume you were Sandra. Current data indicates that 92% of employers check their worker's social media. More specifically, 67% say they use social media sites to research potential job candidates. 54% of companies have eliminated a candidate based on their social media feeds alone. Do you think it would be fair to consider Stan's social media postings as part of your employee selection process? Should you seek legal counsel prior to deciding? Explain your answer.

Key Terms

Application	Reference (employment)	Flat rate (income tax)
At-will (employment)	Negligent hiring	Graduated rate
Good cause	Negligent retention	(income tax)
Validity (test)	Employment agreement	
Reliability (test)	Job offer letter	

Operator's 10-Point Tactics for Success Checklist

Evaluate your need for, and the current status of, each of the following operational tactics. For those tactics you think are important, but not yet in place, develop an action plan for its implementation including who will be responsible for the tactic's completion and the target date by which it should be completed.

Tactic	Don't Agree (Not Done)	Agree (Done)	Agree (Not Done)	If Not Done	
				Who Is Responsible?	Target Completion Date
1) Operator recognizes the importance to success of selecting qualified team members from a pool of job applicants.	——	——	——		
2) Operator understands that every candidate for a job must complete a standardized (uniform) application that must be kept on file for the legally proscribed length of time.	——	——	——		
3) Operator can summarize the legal concept of "employment-at-will."	——	——	——		
4) Operator recognizes that, if interviews are improperly performed, significant legal liability can result.	——	——	——		
5) Operator recognizes that questions asked of applicants during an interview must focus on the applicant's job skills and nothing else.	——	——	——		
6) Operator can list and describe the three main types of pre-employment testing.	——	——	——		

Tactic	Don't Agree (Not Done)	Agree (Done)	Agree (Not Done)	If Not Done	
				Who Is Responsible?	Target Completion Date
7) Operator understands the advantages and potential drawbacks associated with utilizing background checks and references in the new employee selection process.	———	———	———		
8) Operator can explain the significance of understanding negligent hiring and negligent retention.	———	———	———		
9) Operator can list the components of a legally acceptable job offer letter.	———	———	———		
10) Operator recognizes that federal, state, and (sometimes) local documentation must be properly obtained before a newly selected employee can officially begin working.	———	———	———		

7

Orienting and Onboarding New Team Members

What You Will Learn

1) The New Team Member Adaptation Process
2) The Goals of Formal Orientation Programs
3) How to Create a Formal Orientation Program
4) The Importance of New Team Member Onboarding

Operator's Brief

In this chapter, you will learn about three important concepts that affect the introduction of newly selected team members to your foodservice operation. These are:

1) Employee adaptation
2) Orientation
3) Onboarding

All new employees go through an adaptation period by which they learn about their new job and their fellow employees at all organizational levels. The adaptation process takes place regardless of whether you had planned to take specific steps to shape your new team members' initial experiences, or if you did not plan to do so. In this chapter, you will learn about the adaptation process and how you can influence it to optimize your new team members' initial experiences on the job.

Formal orientation programs are often a carefully developed agenda of activities and content designed to train and teach new employees about their roles and your company's important policies. In this chapter, you will learn about the goals of an orientation program and the impact of orientation programs on your new and existing team member, and your guests.

In this chapter, you will learn that there are seven key elements that must be present in any team member's orientation program that are needed for you to provide needed information and address potential legal issues related to this topic. Each of these elements and their importance is addressed in detail in the chapter.

Finally, in this chapter, you will learn about onboarding. Employee onboarding refers to all of those activities including your orientation program that shape your new team members' initial and ongoing perceptions of you, your operation, and their role in your organization.

The chapter concludes with suggested onboarding activities that may be planned for a new team member's first few days on the job. Each incorporates the onboarding philosophy that emphasizes the welcoming of a new team member for the long-term benefit of both the team member and your operation.

CHAPTER OUTLINE

Importance of New Team Member Adaptation
    The Adaptation Process
    Steps in the Adaptation Process
Formal Orientation Programs
    The Goals of Orientation Programs
    The Impact of Orientation Programs on New Team Members
    The Impact of Orientation Programs on Existing Team Members
    The Impact of Orientation Programs on Guests
Key Elements of Orientation Programs
    Compliance with Government Regulations
    Communication of the Operation's Mission And Culture
    Communication of All Employee Benefits
    Communication of Critical Policies
    Introduction to the Operation and Other Team Members
    Detailed Explanation of the New Team Member's Job Duties
    Documentation
Successful Onboarding of New Team Members

Importance of New Team Member Adaptation

After foodservice operators have utilized appropriate recruitment and selection techniques to select a new team member, their next important task is to provide that team member with a proper **orientation**.

In some other publications addressing new worker orientation, the beginning information is

Key Term

Orientation (employee): The process of providing basic information about a foodservice operation that every new employee within the facility should know.

typically organized from the perspective of the employer. Issues such as what should be covered in an orientation program and how that information is best communicated are addressed.

Experienced foodservice operators, however, recognize that *prior* to considering their orientation programs, they must first consider their new team members' needs and desires rather than only their own interests. The reason: Addressing employee needs first is the best way to create an operator's orientation program because, in large part, the process focuses on the organization's new members.

The Adaptation Process

Foodservice operators have an important responsibility to help their new team members learn about and become comfortable working in their new jobs. Whether it is carefully planned or just "happens," all newly employed team members go through an **adaptation** process.

Effective foodservice operators realize that their efforts to meet employees' needs and to reduce turnover rates begin the moment new employees are selected. They understand that

Key Term

Adaptation (to an organization): The process by which new employees learn the values of and "what it is like" to work for a hospitality organization during their initial job experiences.

new staff members are anxious and perhaps even stressed because they do not fully understand specific job expectations or how their performance will be judged. They also are likely to be uncertain about relationships with supervisors and peers, about whether there will be unexpected work tasks, and if there will be unanticipated physical and/or mental challenges. These are among the concerns that operators should address in their earliest interactions with new team members.

For employees to work effectively, they must know what to do and how to perform their assigned job tasks properly. These concerns should begin to be addressed during the recruitment process as potentially new staff learn about job responsibilities that are also addressed in training programs (see Chapter 7) that begin after orientation concludes. However, new team members will see, hear, and experience many things as they begin work that set the context for the more formal experiences that follow. Contrast, for example, two greetings that might accompany the introduction of a new employee to experienced peers: "So glad you're here; welcome to the team," and "Hey, we really need help; hope you stay here longer than the last guy!"

In short, the new employees are looking for initial experiences that are in line with—not contrary to—factors and information provided before position acceptance that were designed, at least in part, to "sell" the new position to the applicant. While a foodservice operator cannot write the script for what a current employee says to a new employee, the operator's history of actions that impact the work environment must be easily and quickly seen as new team members adapt to their new jobs.

The cleanliness of workstations, conversations of employees between themselves and guests, and behaviors of employees that represent their work attitudes will be observed by and influence the attitudes and behaviors of newly hired team members.

New employees want to be accepted by their peers and to quickly become contributing members of their work teams. While this socialization process takes time, it begins as workers are initially put at ease, and as they are involved in hospitable interactions with their peers.

Effective operators know that most new workers want to "fit in" with their peers and become effective team members, rather than advocates of the "them versus me" culture that exists in some operations. These operators have a significant influence over the attitudes and actions of staff members. The precedence they have set with their employers and their ongoing interactions impact current staff members' interactions with new employees.

Steps in the Adaptation Process

Regardless of whether new staff are exposed to a carefully planned adaptation process, several steps are typically involved in the process by which new employees adapt to their organization and the employees within it. Figure 7.1 provides an overview of the new employee adaptation process.

Note: Ideally, foodservice operators will have planned a systematic adaptation program that is logical and focused on the new employee. However, in the absence of an exemplary program such as that shown in Figure 7.1, new staff members will still learn uncontrolled perspectives of the new operation as part of the improperly planned adaptation program to which they were exposed.

✓ *Step 1*—When new employees are selected, they have basic perceptions and attitudes about the work and the organization. These are probably based on factors including (1) information learned during the employment interview,

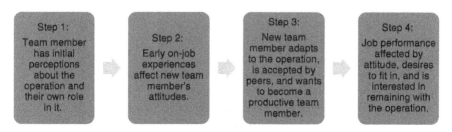

Figure 7.1 Steps in the Adaptation Process

(2) advertising messages (if the new employee has experienced the company's advertising messages), (3) previous experience, if any, as a guest in the operation, and (4) feedback about the property from others in the community, including family and friends and current or former employees.

✓ *Step 2*—Early on-job experiences including those related to adaptation and training may reinforce initial perceptions (Step 1) or they may prove them to be less than accurate. Some apprehension is typical, however, if there is a significant difference between what new employees perceived (Step 1) and what they actually experience (Step 2).

New team members must either make significant changes in perceptions and expectations or, perhaps too frequently, new employees are likely to become discontented and (eventually) part of the organization's turnover statistics. This is especially so when the new employee desires to work for an organization in a position that meets initial expectations (Step 1) and/or when the staff member has other employment opportunities.

✓ *Step 3*—Team members who begin to recognize and accept the culture of the operation and who want to become cooperating members of a work team will likely be accepted by their peers. They will then want to become contributing members of the organization.

✓ *Step 4*—At this point, perhaps the most difficult challenge has been accomplished. The new staff member has a positive attitude about the operation and is willing to learn about and contribute to it. The initial adaptation and training activities enable new employees to perform work that meets quality and quantity standards. Successful performance reinforces the employee's attitude about the organization, and they begin to experience and relate to cultural norms, encouraging retention rather than turnover.

What Would You Do? 7.1

"You quit?" asked Jodi Guild.

"I quit," replied David.

"But today was your first day. You didn't even work a full day?" said Jodi.

"I actually was only there for about an hour and a half," said David.

David and Jodi were both college students at State University. David had accepted a job in a restaurant only two blocks from the residence hall where he and Jodi had their rooms.

David had been excited, three days before, to tell Jodi that he had accepted a job working in the dish room at the restaurant. Because it was so close, it would make it easy for him to get back and forth to work, so his commuting time would be next to zero. It was one of the major things that attracted him to the job.

While the job paid only minimum wage, David knew it would still help with his tuition payments and reduce his reliance on student loans.

"Well, what happened?" asked Jodi.

"I got there at 10:00 o'clock this morning like they told me to. I filled out some forms, and within 15 minutes I was introduced to J.J. Johnson. She was a very nice lady, about 50 I would guess, and she was supposed to show me how they did pots and pans. That was the station I was assigned to."

"O.K., sounds reasonable," said Jodi.

"Well, you would think so," replied David "They wash their pots in a three-compartment sink. The third sink is the final rinse sink. What was amazing to me was that the water temperature in that sink was supposed to be kept at 170°. J.J. told me that was the "code," whatever that means. Anyway, then she told me the pots had to stay submerged for at least 30 seconds, and then you reach in and take the pots out".

"And the problem was...?" asked Jodi.

"And the problem was that I couldn't do it. I tried. A couple times. I couldn't grab the pots. They were just too hot. When I got back to my dorm room, I looked it up. Water temperatures above 120° will scald you, . . . and pretty seriously!"

"What did J.J. say?" asked Jodi.

"When I asked J.J. about it about it, she showed me her forearms. They were bright red, and then she told me I'd get used to it like she did. I thought about it awhile, but it was pretty easy for me to decide. I didn't want to get used to it," said David.

"So, you just left," said Jodi.

"That's right. I just left," said David "I didn't want to cause any problems."

Assume you were the operator of this foodservice facility. Why did David quit? What could you have done to prevent that? Do you think David actually wanted to quit his new job? Explain your answer.

Formal Orientation Programs

Those foodservice operators who fully understand the employee adaptation process are in the best position to begin to design a formal (and effective) **orientation program** that will positively impact the adaptation process.

It is critical that an effective orientation program be planned and consistently implemented because it impacts the initial and on-going

Key Term

Orientation (program):
A carefully developed agenda of activities and content designed to train and enlighten new employees about their roles and company policies.

relationship between the organization and its newly hired staff members. As they begin the process of developing their formal orientation programs foodservice operators should consider four important areas:

1) The Goals of Orientation Programs
2) The Impact of Orientation Programs on New Team Members
3) The Impact of Orientation Programs on Existing Team Members
4) The Impact of Orientation Programs on Guests

The Goals of Orientation Programs

A foodservice formal operator's orientation program is designed to help achieve a number of goals including:

✓ *Providing an overview of the organization*—Many newly employed team members want to know their employer's history, size (number of locations and staff members, for example), and the products and services it provides. They should also learn about the business results that their new employer is attempting to achieve. New workers will want to know how the organization adds value for its guests, to themselves, and to the organization's owners. An orientation program provides a way to communicate an operation's mission statement (see Chapter 1) explaining what the organization wants to accomplish and how it intends to do so.

✓ *Describing the new team member's role*—In most cases, a new staff member will benefit from seeing an **organizational chart** that shows all a foodservice operation's positions (including their own), and the reporting relationships between them.

Key Term

Organizational chart:
A diagram that visually conveys a company's internal structure and its reporting relationships.

An organizational chart represents an operation's current structure, and it lets new employees know about promotion opportunities that might be available if they perform well.

Find Out More

Some foodservice operators, and especially those whose businesses are small, may feel that there is no need for an organizational chart. However, an organizational chart is an excellent visual guide to the key positions in a business. Therefore, it should be shared with new employees and, if the operator desires, the names of the individuals holding the positions.

Fortunately, it is extremely easy to create an organizational chart for a foodservice operation. Various websites offer operators the ability to create their own organizational chart at no charge. In most cases, such a chart can be created and printed in less than five minutes.

To see how easy it is, enter "organizational chart templates for restaurants" in your favorite search engine and review the results.

✓ *Explaining policies, rules, and other information*— New team members will want to know general work guidelines, including their days and hours of work, uniform requirements, break times, auto parking, and other similar information to help them feel more comfortable.

✓ *Outlining specific expectations*—Topics in an orientation program should include responsibilities of the employer to the new employee and, in turn, the responsibilities of new team members to the employer will also be addressed.

✓ *Motivating new team members*—The enthusiasm and excitement exhibited by those providing orientation experiences to new workers are important. Orientation helps to establish a solid foundation for a positive relationship between the organization, its managers and supervisors, and new team members.

Taken together, the benefits of effective orientation programs can eliminate confusion, heighten a new staff member's enthusiasm, create favorable attitudes, and, in general, help make a positive first impression.

The Impact of Orientation Programs on New Team Members

The number one priority goal of an orientation program should be to help new team members feel comfortable in their positions and to provide helpful information that enables them to maximize their contribution to the organization.

The program should also be planned to reduce new workers' possible anxiety and stress by addressing the types of questions and concerns they might typically have as they begin their new jobs. Examples of questions and concerns include:

✓ Who is my immediate supervisor/boss?
✓ What are my duties?
✓ What are my rights?
✓ If I do a good job, what advancement opportunities (and pay increases!) might exist for me?

Orientation programs are most effective when they help to establish a relationship between the new staff member and the organization that is based on both parties helping each other to attain important goals. The process begins by ensuring that orientation programs directly address key concerns and questions that will be "top of the mind" for all new workers.

The Impact of Orientation Programs on Existing Team Members

For most foodservice operators, an important portion of an orientation program includes having new team members meet other members of their work team. When this component is in place, existing team members will feel more comfortable working alongside new employees because they have gotten to know them on somewhat of a personal level.

Existing employees also know that the new workers have been informed of their responsibilities in the job and that everyone is working from the same "playbook." Formal employee-to-employee introductions can also let existing workers know where they may be of most assistance in helping new workers acclimate to their jobs. This can provide existing workers with a sense of importance and helpfulness that serves to motivate them and, as well, can assist the new worker.

The Impact of Orientation Programs on Guests

Properly oriented employees do a better job serving guests. New team members who have been informed about an organization's mission statement and goals are in a better position to deal with guests who may be angry, frustrated, or concerned about a problem with a menu item or a service issue.

While in some cases the new employee may not be able to directly resolve the issue, properly oriented new employees will know exactly who to go to for assistance with the guest's specific concern.

Studies have consistently shown that employees who are more comfortable in their own jobs are more pro-active in dealing with guests. A properly oriented team member feels more confident assisting guests with questions or solving issues. Even in uncertain scenarios, comfort in their positions can build their guests' trust in the worker, and the new worker's trust in their employer.

Key Elements of Orientation Programs

The specific information included in a foodservice operator's formal orientation program varies by property. Some foodservice operators include materials used for their orientation programs in an **orientation kit**.

Examples of items that can be included in an orientation kit include a copy of the current

Key Term

Orientation kit: A package of written materials given to new employees to supplement the oral information provided during an orientation session.

organization chart, employee handbooks, and copies of employee performance appraisal forms and procedures. Other examples can include federal, state, and/or local tax materials, a diagram of the property layout (if it is large), and accident prevention and emergency procedural guidelines.

Regardless of its specific content, an effective new employee orientation program should address key elements:

1) Compliance with government regulations
2) Communication of the operation's mission and culture
3) Communication of all employee benefits
4) Communication of critical policies
5) Introduction to the operation and other team members
6) Detailed explanation of the new team member's job duties
7) Documentation

Compliance with Government Regulations

As the beginning of the employment relationship begins, employers must work with employees to have them complete multiple government-mandated forms. As addressed in the previous chapter, these can include federal and state tax forms and I-9 forms.

While such documents should technically be completed before a new team member's first day of work, the orientation program gives the employer the chance to confirm that all proper documentation is in place. This can include age verification documentation (for minors) if this was not obtained and retained during the selection process.

It is important to note that employers can face significant penalties, including monetary fines, for not properly obtaining this documentation.

Communication of the Operation's Mission and Culture

In the rush to complete the delivery of the orientation program, some foodservice operators overlook the need to emphasize their operation's mission and culture. Taking the time needed to communicate these messages to new team members is essential to ensuring new employees are aware of their employer's business goals.

Where possible, some foodservice operators include direct contact with their operation's owners or founders. Such an introduction can help to reinforce the operation's culture. When that is not possible, a written note of welcome or introduction from the owner(s) of an operation can send a powerful welcoming message to new employees.

Communication of All Employee Benefits

In nearly all cases, a new team member's rate of pay will have been established during the interview and job offer process. However, in some organizations the cost of employee benefits can approach up to 50% of new workers wage or salary. While it is, of course, important to tell new team members the operation's rules and policies that could result in discipline or even termination, foodservice operators also need to sell the **employee benefits** they provide to new team members.

Key Term

Employee benefits: Any form of rewards or compensation provided to employees in addition to their base salaries and wages. Benefits offered to employees may be legally mandatory or provided voluntary.

Employee benefits consist of costs associated with, or allocated to, payroll, and they must be paid by employers. Examples include items such as payroll taxes (e.g., mandated contributions to Social Security and worker's compensation plans) and voluntary benefit programs offered by the operation.

The costs of these mandatory and voluntary benefit programs can be significant, and they include costs such as:

✓ Performance bonuses
✓ Health, dental, and vision insurance
✓ Life insurance
✓ Long-term disability insurance
✓ Employee meals
✓ Sick leave
✓ Paid holidays
✓ Paid vacation

Fully informing new team members about all of these and/or other benefits can help the employment relationship begin on a positive note.

Communication of Critical Policies

Numerous laws regulate the workplace, and thoughtful operators will want to create a paper trail showing that their employees were informed about the employer's policies and commitment to compliance with these laws. To cite just one example, in the United States the federal government does not currently require foodservice operators to administer sexual harassment prevention training to its workers. However, recall from Chapter 3 that the regulations issued by the Equal

Employment Opportunity Commission (EEOC) regarding sexual harassment provide that:

> *[a]n employer should take all steps necessary to prevent sexual harassment from occurring, such as affirmatively raising the subject, expressing strong disapproval, developing appropriate sanctions, <u>informing employees of their right to raise and how to raise the issue of harassment under Title VII,</u> and developing methods to sensitize all concerned.*[1]

Including information on harassment in a new employee's orientation kit is a good way to demonstrate compliance with the <u>*informing employees of their right to raise and how to raise the issue of harassment under Title VII*</u> recommendation addressed above.

Individual states too, may have their own regulations that can impact employee orientation programs. For example, effective January 1, 2021:

> *every California employer having five or more employees shall provide at least two hours of classroom or other effective interactive training and education regarding sexual harassment to all supervisory employees and at least one hour of classroom or other effective interactive training and education regarding sexual harassment to all nonsupervisory employees. . . . In addition, beginning January 1, 2020, for seasonal, temporary, or other employees that are hired to work for less than six months, an employer shall provide training within 30 calendar days after the hire date or within 100 hours worked, whichever occurs first.*[2]

In some cases, even local legislation can impact the content of an orientation program. As a result, foodservice operators seeking to communicate information about critical policies should always consider including information about one or more of the following topics in their orientation kits:

✓ Equal employment opportunity statement
✓ Zero tolerance harassment policy
✓ Family and medical leave (or similar state leave laws)
✓ Reasonable accommodations
✓ Limits on electronic communications use
✓ Social media/blogging
✓ Confidential information

---

1 https://www.ecfr.gov/current/title-29/subtitle-B/chapter-XIV/part-1604/section-1604.11. Retrieved May 5, 2023.
2 https://www.cnbc.com/2020/06/15/supreme-court-rules-workers-cant-be-fired-for-being-gay-or-transgender.html. Retrieved October 23, 2023.

✓ Safety and health-related issues
✓ Continuation of medical insurance benefits during leave of absence (s)
✓ Ongoing drug testing (if applicable)
✓ Firearms or other weapons in the workplace
✓ Workplace violence
✓ Proper operation of motor vehicles
✓ Recreation or medical drug use

Find Out More

The sale and possession of firearms is an emotional and often controversial issue in the United States. In some states, the right to possess firearms is expanding. In other states, efforts are being made to limit the sale and purchase of guns.

It's important to recognize, however, that currently in every state, the right of a business owner to prevent members of the public from bringing firearms into their businesses has been preserved. But, in some cases, it is preserved only if gun owners have been given oral or written notice that entry with a firearm is prohibited. This written notice applies to the employees of the business as well, and such notice may be an important part of a foodservice operator's orientation kit.

The laws regarding possession of guns in the United States are rapidly changing, and thoughtful foodservice operators will want to keep up with those changes. To learn more about the rights of foodservice operators to control the presence of firearms carried by guests or employees on their premises, and in the manner they feel is appropriate, enter "the rights of employers in 'right to carry' gun states" in your favorite search engine and review the results.

While the specific content of a foodservice operation's orientation program varies, Figure 7.2 lists some suggested topics that many new employees will want to know about, and those that are applicable should be considered for potential inclusion in an operator's formal orientation program.

In addition to understanding a foodservice operation's policies, new team members also need to be informed about the procedures for reporting violations of these policies, such as how to report alleged violations of the no harassment policy and safety rules.

Experienced foodservice operators know that a properly prepared employee handbook (see Chapter 4) is an important orientation tool because it will directly address an operator's policies and employment practices. It is typically distributed and discussed as part of the general orientation process for all new employees or, more recently, employees undergoing orientation are given access to the handbook with an online link to a secure and protected (employees only) site.

Absenteeism Rules	Drug Testing	Parking	Smoking Areas
Accrual of vacation	Emergency plans	Performance standards	Termination
Appearance and grooming	Jury duty	Personal leave requests	Uniforms
Appraisal systems	Meal programs/ meal charges	Phone use (personal)	Vacation scheduling
Breaks	Multiple employment policies	Probationary periods	Weapons
Disciplinary process	Overtime pay	Safety/security issues	Workplace violence
Dress code	Paid holidays	Schedule posting/ modifying	Zero tolerance harassment policy

Figure 7.2   Potential Orientation Program Topics to Address

Technology at Work

A digital employee handbook can be used to communicate a foodservice operation's critical policies, procedures, and working practices.

There are a number of advantages to utilizing a digital employee handbook. These include:

1) Reduced paperwork: A digital employee handbook is easy for workers to access. It also costs less to up-date a digital handbook as compared to paper-based handbooks.
2) Central storage location: Critical information is in one spot known to all employees.
3) Usage tracking: Modern digital employee handbooks allow foodservice operators to know whether an employee has accessed handbook information because an electronic record is kept of each employee's access.

Today, a number of organizations exist to assist businesses in developing digital handbooks that can be stored safely and securely online.

Information in the handbooks can be protected from the general public using specified employee IDs and passwords. To learn more about how foodservice operators can convert employee handbooks from hard copy to digital communication tools enter, "how to create a digital employee handbook" in your favorite search engine and view the results.

Introduction to the Operation and Other Team Members

In larger operations, it may be helpful to include a tour of the business to ensure new team members know how to find their way around. Foodservice operators should also take some time in their orientation programs to introduce new team members to as many other team members as is reasonably possible. Also, it is always helpful for new team members to know the roles of their co-members and how their own job relates and connects to the overall objectives of the team. In part, this responsibility can be enabled by a discussion of a current organization chart for the property.

Detailed Explanation of the New Team Member's Job Duties

While it is most likely that the job duties related to a specific position will have been addressed during a job applicant's interview process, the orientation program gives foodservice operators the opportunity to explain the role of a new team member very specifically. Doing so properly will most often include supplying details about:

✓ The factors used to evaluate job success
✓ Who will evaluate the new team member's performance
✓ Resources that are available to help team members in their first days of working for the operation
✓ How the team members' job impacts the overall success of the foodservice operations

In many cases, the foodservice operations identify a specific current employee who will serve as a coach and guide to a new team member (see Chapter 8).

Documentation

Documenting the activities undertaken and the information delivered an orientation program can be critical if a foodservice operator must prove their compliance with applicable laws. Also, new employees should be able to help address a variety of potential issues including claims for unemployment compensation, citations for food safety rule violations, and charges of discrimination, harassment, or **retaliation**.

Foodservice operations that sell alcoholic beverages often begin their emphasis on responsible service during their orientation programs.

Key Term

Retaliation: Any act that results in an employer punishing an employee for advocating personal rights to be free from employment discrimination. These include a discriminatory workplace culture, violations of laws intended to protect health and safety, or acting as a whistleblower.

Note: This issue relates beyond food and beverage personnel to others who meet with guests in a property that offers alcoholic beverage service. The provision of this information helps establish a priority for concern. Also, it also establishes a record of consistent and ongoing emphasis that can help to defend the organization if alcoholic beverage service-related lawsuits arise in the future.

Some management specialists might argue that formal orientations should be designed to appeal to all employees equally (not just selected subsets of workers). However, the reality is that some specific employees are interested in specific information because of their own unique circumstances. Thoughtful foodservice operators will be aware of these special concerns and address them as appropriate.

For example, consider that, in the past, food and beverage menus would most typically be printed on paper and in a format that was intended to be easily handed to, and then read by, guests. While these menus are still prevalent, increasingly some foodservice operators create a **digital menu** that allows them to easily change menu items and prices without the printing costs associated with hard copy menus.

In fact, some operations create no printed menus at all, and they prefer to instruct guests in the use of a **QR code** and their smart devices to access the operations' menus.

For most younger and technologically savvy employees who join a foodservice operation's service team, the use of QR codes is routine. Assume, however, that an older employee has been hired to work as a server. While some older workers are also very tech savvy, others may be totally unfamiliar with QR codes. They may not have experienced them in their previous foodservice jobs and will need careful coaching to learn a system used by guests who may also be unfamiliar with proper procedures. It is obvious that problems will be created unless the new employees receive and practice procedures from other employees who are very familiar with a relatively new ordering system.

Key Term

Digital menu: An integrated system that uses hardware and software to display an operation's menu on an electronic screen; also commonly referred to as a digital display menu or digital menu board.

Key Term

QR code: A QR (quick response) code is a machine-readable bar code that, when read by the proper smart device, allows foodservice guests to view an online menu. Guests can also be re-directed to an online ordering website or an app that allows them to order and/or pay for their meal without having to interact directly with a staff member.

In a similar manner, many foodservice workers, female and male, have primary responsibility for child or parental care. Many of these staff members would benefit from knowing what steps, if any, a foodservice operator has taken to secure temporary care services, so an employee need not call in sick or be late for work.

Note: Not all employees have these child or parental care responsibilities, and foodservice operators should carefully review their orientation information to identify specific content applicable to some, but not all, new employees.

As a final example of customized orientation programs, some new employees with physical disabilities may need special accommodation of their work schedules. New team members with disabilities may require flexible schedules due to:

✓ Medical treatment related to the disability
✓ Repair of a prosthesis or equipment
✓ Temporary adverse conditions in the work environment (for example, an air-conditioning breakdown causing temperature above 85° could seriously harm the condition of a person with multiple sclerosis)

Depending on the nature of the work assignment and operational requirements, changes to work schedules and hours may be a reasonable accommodation (see Chapter 2) as long as it does not result in an **undue hardship**.

Thoughtful foodservice operators designing their orientation programs consider whether reasonable information related to medically mandated schedule adjustments should be privately discussed with an individual new team member during their orientation.

New team member orientation programs can be tailored to the needs of each foodservice operation. However, addressing the seven essential elements of an employee orientation program presented above, foodservice operators can maximize their chances for properly welcoming new team members. This, in turn, can help to avoid excessive turnover rates and future labor and employment-related legal claims.

Key Term

Undue hardship: Action requiring significant difficulty or expense when considered in light of a number of factors. These factors include the nature and cost of the accommodation in relation to the size, resources, nature, and structure of the employer's operation.

Technology at Work

There are many cases in which foodservice operators seek information about developing effective orientation program. The results of their internet search are often information suggesting the ideal orientation setting involves meeting with employees in a single room with an orientation leader who carefully explains key information.

The reality for most foodservice operators, however, is that new employees are *not* typically oriented as a group. Instead, and especially in small foodservice operations, new employees are usually oriented one at a time rather than in large groups. The reason: you have learned that it may be ineffective or even unlawful for a foodservice operator to wait to orient a new employee until a large group of new team members has been identified. As a result, orientation sessions often take place with a single new team member.

Orientation sessions can require a significant amount of time and increasing numbers of foodservice operators now create orientation videos that address many of the key issues about which team members learn. Fortunately, todays' foodservice operators can create an orientation video addressing some or most of the information about which new team members must be informed and new team members can watch these videos at any time.

Creating these videos is increasingly easy using a smartphone. For more information about how to use smart devices to create videos appropriate for use in an orientation program, enter "how to make and edit videos using a smartphone" in your favorite search engine and view the results.

Successful Onboarding of New Team Members

While the proper recruitment, selection, and orientation of new employees is critically important, foodservice operators must also consider a new team member's entire **onboarding** experience. Note: Onboarding refers to the process by which a new employee is welcomed and integrated into a foodservice operation. Some observers believe the process begins with the first contacts a person seeking employment has with an operation.

As new team members participate in their initial work experiences, they are looking for reinforcement that their decision to join a foodservice operation was a good one. What operators do (and don't do!) in the initial employment period makes a significant and lasting impact on new team members' long-term perceptions. The onboarding process, which actually begins with recruitment and selection, should continue as the new employee begins work.

Key Term

Onboarding: The process in which new team members are integrated into an organization. It includes activities that allow new employees to complete an initial new-hire orientation process, as well as learn about the organization and its structure, mission, and values.

Typically, the first day's activities should be limited to what a new employee can reasonably be expected to learn and do on a first shift. The actual orientation program may or may not be held on the first day; however, it is important that new employees be made to feel genuinely comfortable and happy about their new employment decision.

Below are suggestions for onboarding activities that may be planned for the employee's first day on the job. Each incorporates the onboarding philosophy that emphasizes the welcoming of a new team member.

✓ Providing existing staff with a summary of the new employee's background before the new team member begins work. Encouraging current staff to genuinely welcome the new employee, explain how their roles interact with that of the new hire, and discuss how they might work together.

✓ Assigning an experienced employee to serve as a short-term mentor to support and advise the new team member. The designated employee can then be an immediate resource for questions and provide useful information as the new employee becomes acclimated to the job and the operation.

✓ Setting up the new employee's workstation, if applicable, and ensuring their assigned employee lockers have been cleaned out (and is clean!) and are readily accessible. Note that when these types of basic welcoming tasks have been overlooked, the new employee may consider the lack of planning and preparation to be an early sign of an uncaring employer.

✓ It will be much more than a "nice touch" if the owner/senior manager of the operation personally meets and greets the new employee on their first day. Even if a peer supporter is assigned, the employee's immediate supervisor should attempt to visit informally with the employee.

It is important to recognize that many people form impressions of an operation before they begin to work in it. Consider, for example, an "employer of choice" (see Chapter 1) for whom many within the community want to work, and the job application rate is high when there are openings. In contrast, consider an "employer of last resort" for whom people work only until they can find preferred employment, and the lessons are clear. The hiring of each team member gives the foodservice operator the chance to onboard new team members in a manner that contributes to the potential of becoming an area's employer of choice.

There is little doubt that early experiences on the job can positively or negatively affect the adaptation process of all new team members. Foodservice operators utilizing a well-planned orientation program can apply practical onboarding activities in the first few days of employment. This, in turn, can make a tremendous impact on their teams' effectiveness and the longevity of each new team member.

Of all the activities that most impact new team members and their onboarding experiences, the training necessary to be successful in their jobs is certainly among the most important. Unfortunately, in some foodservice operations, new (and current) employee training is extremely limited, or even non-existent. The critical nature of team member training, and how to do it well, is so important to the long-term success of a foodservice operation that it will be the sole topic of the next chapter.

What Would You Do? 7.2

"Well, I think we need to spend more time in our orientation session explaining why it's so important they arrive on time. Even though they have early schedules, if they don't get here when they're supposed to, we can't get everything ready for our customers by the time we open," said Patti Stewart.

"That has been a problem," said Bess Haley.

Patti and Bess were the co-owners and operators of Best Baked Breads, the "Triple B," as they called it, featured high-end baked products, and it specialized in artisan breads. The operation was open seven days a week from 6:00 AM to 3:00 PM.

Patty and Bess were reviewing their new employee orientation program. Like many foodservice operators, they were in a tight labor market. They had experienced the situation of hiring what they perceived to be good employees but, in many cases, attendance was sporadic and some employees, especially the younger ones, often arrived later than their 5:00 AM starting shifts.

"Well, I don't think we'll have any problem explaining why it's important to us that they arrive on time. But maybe the better question is, should we spend more of our time telling them what's important to us or addressing what's important to them?"

Assume you were Patti and Bess. Do you think attendance and punctuality-related issues such as the one they are addressing are best addressed in the recruiting process, the selection process, the orientation process, or all three processes? Explain your answer.

Key Terms

Orientation (employee)
Adaptation (to an organization)
Orientation (program)
Organizational chart
Orientation kit
Employee benefits
Retaliation
Digital menu
QR code
Undue hardship
Onboarding

Operator's 10-Point Tactics for Success Checklist

Evaluate your need for, and the current status of, each of the following operational tactics. For those tactics you think are important, but not yet in place, develop an action plan for its implementation including who will be responsible for the tactic's completion and the target date by which it should be completed.

Tactic	Don't Agree (Not Done)	Agree (Done)	Agree (Not Done)	If Not Done	
				Who Is Responsible?	Target Completion Date
1) Operator recognizes that all new team members go through a clearly defined adaptation process as they learn their unique roles in a foodservice operation.	——	——	——		
2) Operator can state the goals of a new employee orientation program.	——	——	——		
3) Operator understands that a formal orientation program directly impacts their new team members, existing team members, and guests.	——	——	——		
4) Operator recognizes that compliance with government regulations is one key element of a formal orientation program.	——	——	——		
5) Operator recognizes that communication of their operation's mission and culture is one key element of a formal orientation program.	——	——	——		
6) Operator recognizes that the communication of employee benefits and critical policies are two key elements of a formal orientation program.	——	——	——		

Tactic	Don't Agree (Not Done)	Agree (Done)	Agree (Not Done)	If Not Done	
				Who Is Responsible?	Target Completion Date
7) Operator recognizes that an introduction to the operation and other team members is a key element of a formal orientation program.	——	——	——		
8) Operator recognizes that documentation is a key element of a formal orientation program.	——	——	——		
9) Operator understands the importance of customizing, where applicable, the content of their orientation programs to meet the unique needs of some new team members.	——	——	——		
10) Operator recognizes that onboarding is a continuous process that begins when applicants first apply for work and continues throughout a new team member's first days of employment.	——	——	——		

8

Planning for Employee Training

What You Will Learn

1) The Importance of Proper Training
2) Training Principles
3) Characteristics of Effective Trainers
4) How to Create a Successful Training Program
5) How Online Guest Reviews Impact Training Plans

Operator's Brief

In this chapter, you will learn that to be effective, you must carefully plan for your operation's employee training program. Many benefits can accrue to your foodservice operation when you do so, especially when you overcome some training obstacles and myths.

There are a number of training principles that you should know as you plan training programs and, in this chapter, you will learn about some of the most important of these principles.

Experienced foodservice operators know that effective trainers must possess specific personal characteristics to be successful. In this chapter, you will learn about these characteristics and why they are important.

The actual creation of an effective training program is a five-step process, and these steps must occur in the sequence cited below:

✓ Step 1: Identify Training Needs
✓ Step 2: Develop Training Objectives
✓ Step 3: Develop Training Plans
✓ Step 4: Create Training Lessons
✓ Step 5: Develop (and Update) Training Handbooks

The creation of an effective training program begins when you identify your training needs. There are many common sources that can be used for this process, and these are identified in this chapter. After training needs have been developed, reasonable and measurable training objectives can be established.

The development of a comprehensive training plan is important, and in this chapter, you will learn how these plans are created. After a training plan has been created, individual lessons that make up a trainee's specific training program can be developed, and these will vary based on the specific position held by each trainee.

You will also learn that the development of a training handbook is an important process. The reason: This training tool is helpful in minimizing the work required to implement your training plans and in documenting your training efforts.

The chapter concludes with a discussion about technology and the use of information found on user-generated content (UGC) sites as you plan your training programs.

CHAPTER OUTLINE

The Importance of Training
    Benefits of Training
    Training Obstacles and Myths
Training Principles
Characteristics of Effective Trainers
Steps in Creating Effective Training Programs
    Step 1: Identify Training Needs
    Step 2: Develop Training Objectives
    Step 3: Develop Training Plans
    Step 4: Create Training Lessons
    Step 5: Develop (and Update) Training Handbooks
Technology and User-Generated Content (UGC) Scores

The Importance of Training

Foodservice operations are labor-intensive. Technology has significantly affected how foodservice operations use the internet for marketing, and software advances allow foodservice operators to collect and analyze guest data more efficiently. However, technology generally has not affected the number of employees required to produce and serve the menu items foodservice guests want to buy.

Whether the operation is a food truck or a high-end table service restaurant, newly hired team members must acquire the knowledge and skills needed to excel in their positions. Also, their more experienced peers must often obtain new knowledge and skills to keep up with changes in work procedures, menu items, and services offered to guests. As a result, proper **training** is critical for both an operation's new and existing team members.

Key Term

Training: The process of developing a staff member's knowledge, skills, and attitudes necessary to perform tasks required for a position.

Benefits of Training

There are numerous benefits to a foodservice business when a foodservice operator makes staff training a major priority:

1) *Improved worker performance*—Trainees learn knowledge and skills to perform required tasks more effectively and their on-job performance improves.
2) *Reduced operating costs*—Improved job performance helps to reduce errors and rework, and associated costs are reduced. Workers performing their jobs correctly will be more productive and fewer labor hours will be needed.
3) *More satisfied guests*—Training can yield more service-oriented employees who know how to please guests and are motivated to do so.
4) *Fewer operating problems*—Busy operators can focus on priority concerns and will not need to address routine operating problems caused by inadequate training.
5) *Lower employee turnover rates*—Fewer new staff members become necessary as turnover rates (see Chapter 1) decrease. Those who are properly trained and rewarded for successful performance are less likely to leave and operators have less need to recruit new employees.
6) *Higher levels of work quality*—Effective training identifies quality standards that define acceptable product and service outputs. Trained employees are more interested in operating equipment correctly, in preparing menu items the "right" way, and in properly interacting with guests.
7) *Improved morale*—The morale of staff members generally improves when they are impressed with their employer's ongoing commitment to provide training that allows advancement to more responsible and higher-paying positions. High levels of morale, in turn, are important to the success of the training process.

For example, assume an operator's training program has a goal of addressing food safety concerns, and it is a priority for food production personnel.

Further, assume also that the training addresses important information that must be known about this topic. If a cook has a high level of morale and wants to use the safe food practices they have learned, the training is successful. If, however, due to low morale and/or other factors, a trainee is not motivated to follow safe food-handling practices, the training will not be successful. This problem can occur even though the "why" and "how" aspects of training were properly communicated to and learned by the trainees.

8) *Easier to recruit new staff*—Satisfied staff members tell their family and friends about their positive work experiences, and their contacts may become candidates for position vacancies that may arise. Foodservice operations that emphasize training can evolve into "employers of choice" that provide "first choice" rather than "last chance" employment opportunities.

9) *More professional staff*—Professionals want to do their job as best they can, and this is only possible with appropriate training.

10) *Greater profits*—If guests are more satisfied and revenues increase, and if labor and other operating costs are reduced, there is significant potential for increased profits. In the long run, training must "add value," which should be measured by the difference between the increased profits and the added training costs. While this measurement is not easy, most industry observers believe that, if done correctly, training does not "cost," but rather, it actually "pays."

Training Obstacles and Myths

Despite the many benefits of effective training programs, in some foodservice operations training does not receive sufficient attention. One reason is that there are obstacles and myths (untruths) associated with the implementation of the best training programs.

Training Obstacles
Some obstacles to effective training can include:

✓ *Lack of trainer time*—In most cases, the routine completion of daily tasks keeps foodservice operators and supervisors very busy. Some of these individuals may feel that they do not have the time required to carefully plan and properly deliver effective training programs.

✓ *Lack of trainee time*—In many cases, supplying formal training means that staff members will be attending training sessions rather than completing normally assigned tasks. Especially when staff shortages exist, finding time for trainees (who are already busy workers!) can be a challenge. Unfortunately, sometimes opportunities to schedule training sessions do not receive a priority from both the presenters and the trainees.

✓ *Lack of money*—Foodservice operators initiating training programs incur real costs both in the staff time required to deliver and participate in the programs and in the acquisition of program materials themselves.

✓ *Trainers' insufficient knowledge and skills*—People must be taught how to train just as they must be taught to perform any other unfamiliar task. Unfortunately, formal **"train-the-trainer"** programs are not provided by many foodservice operations.

✓ *Lack of quality resources available for training*—Many operators do not have the time, knowledge, or ability to develop training videos and/or to prepare extensive or sophisticated training resources or training evaluation tools. If these items can't be developed in-house, they may be available **off-the-shelf**.

Key Term

Train-the-trainer (program): A training framework that turns employees into subject matter experts who can effectively teach others.

Key Term

Off-the-shelf (training programs): Generic training materials typically addressing general topics of interest to many trainers that can be used if operation-specific resources have not been developed.

Technology at Work

In most foodservice operations, the training needs of employees are unique to the operation. Since menu items offered vary and because the method of product delivery to guests may differ as well, several types of employee training must address operation-specific information.

In some other cases, however, operators may be able to obtain important and relatively inexpensive off-the-shelf training products developed specifically for the foodservice industry.

For many foodservice operators, some of the most important of these training resources address the issue of safe food handling. Several different entities have created food safety training videos, and these are often offered at no or at a very low cost. In most cases, food safety training videos are superior to materials written by foodservice operators because employees viewing the videos can see safe food handling practices in action. Such videos most often show employees what they should do, as well as what they should *not* do.

To review food safety-related videos and their costs, enter "food handler safety training videos" in your favorite search engine and view the results. *Note*: Among the materials that can be reviewed are those developed by the National Restaurant Association (NRA) and the American Hotel & Lodging Educational Institute (AHLEI).

Resources addressing generic topics such as supervision tactics, sanitation, and food safety can be purchased for a modest cost. However, excellent foodservice operators are creative, and they would never elect to "not train" simply because supplemental resources are unavailable. Rather, they would likely create their own training tools.

✓ *Scheduling conflicts*—When can food servers learn about new tasks or new menu items offered? Operators must often schedule training sessions while at the same time scheduling other employees to perform required work tasks.

✓ *Turnover*—In many foodservice operations, some staff members leave within a few months (or less!) of initial employment. Some operators may think, "Why train employees if they don't remain on the job long enough to use what they have learned or if they become employed by my competition?" In fact, as noted above, effective training can reduce turnover rates, and those operators who do not train employees may be contributing to their own property's unacceptably high turnover rates.

✓ *Insufficient lead time*—In many cases, too little time may pass between a new team member's first day on the job and the time when they must actually begin performing work tasks. Foodservice operators should take care to ensure that a "**warm body syndrome**" is never used as a selection tactic.

Key Term

Warm body syndrome: A selection error that involves hiring the first person who applies for a vacant position.

✓ *Difficulty in maintaining training consistency*—Individual trainers may plan and deliver training activities based only on what they personally "think" staff must know. Unfortunately, a result of this action is that the "what" and "how" of training will likely be inconsistent. Then those who train may begin to think that "We tried to train, and it hasn't worked very well. There must be a better problem resolution alternative than training. What else can we do?"

✓ *Trainer apathy*—There should be reasons for trainers to want to train. Benefits for successful training duties can include special privileges, compensation increases, promotion consideration, educational opportunities, and/or other types of recognition. By contrast, when trainers must assume these duties in addition to other tasks, if they do not receive train-the-trainer training, and/or if there is little (or no) support for training, why should trainers want to do so?

Training Myths

While training obstacles may be present, in most cases they can be overcome. In addition to training obstacles, however, there are also several myths (nontruths) about training, and they include:

✓ *Training is easy*—In fact, when training involves only a trainee "tagging along" with a more experienced staff member, it is easy. However, the lack of planning and the increased possibility that basic training principles will be disregarded increases the likelihood that this form of training will be ineffective.

✓ *Training costs too much*—Foodservice operations with a history of inadequate training yielding unsatisfactory results may not invest in the resources to plan and deliver more effective training. "Been there, done that" is a philosophy that can easily evolve when only the costs, rather than the benefits of effective training, are considered.

✓ *Only new staff need training*—New team members most often do need training, but so do their more experienced peers. Consider, for example, when operating procedures are revised, new menu items are introduced and/or new equipment is purchased.

✓ *There is no time for training*—Many priorities compete for a foodservice operator's time. When they believe their positions are already extremely busy, training is often de-emphasized, and any available time is allocated to other tasks.

Despite the obstacles, and myths associated with, to effective training, the best foodservice operators recognize that only a highly trained staff can serve guests while allowing an operation to meet its financial objectives. As a result, proper training of staff is a necessity rather than a luxury.

Training Principles

Experienced foodservice operators know that "an organization pays for training even if it doesn't offer it." This concept recognizes that developing and delivering high-quality training programs take time and money to do well. In the absence of training, however, time and money are wasted because employees' consistently create errors that result in rework. Also, guests are less likely to receive the proper quality of food items and services that were ordered.

The training needs of a specific foodservice operation can be as unique as the operation itself. An operation may be a portable food truck, commercial restaurant, or nonprofit employee cafeteria. Regardless, however, if one accepts the idea that foodservice operators will, one way or another, pay for training, it makes good business sense to implement effective training that returns benefits exceeding costs.

A first step in the process is to incorporate basic principles into the training program. The following training principles apply to the largest and smallest foodservice operations regardless of location, type of guests served, or financial objectives that are pursued.

✓ ***Training Principle 1:*** *Trainers Must Know How to Train.* In many foodservice operations, a supervisor or existing co-worker serves as the trainer. The supervisor may have been a "good" employee who is promoted and now performs tasks, including training, which were not part of their previous position. However, existing team members do not become effective trainers by "magic" or by default. Instead, they must be taught how to train, and "train-the-trainer" programs are needed to provide the necessary training knowledge and skills to be a successful trainer.

This principle also applies to a new team member's peer who may be assigned to conduct training. If training fails because the trainer doesn't know how to do it properly, the problem rests with the trainer, not with the trainee. Effective training requires more than just one's willingness to do it.

✓ ***Training Principle 2:*** *Trainees Must Want to Learn.* The old expression "You can lead a horse to water, but you can't make it drink" applies here. Trainees must want to learn, and they must recognize the worth of the training. "Because the boss says it is necessary" is not a meaningful reason from the perspectives of most workers. By contrast, the reason, "this training is a step in a career-long professional development program to help you become eligible for promotion" or "learning to do this right will help increase your tips" may be of interest to many trainees.

✓ ***Training Principle 3:*** *Training Must Focus on Real Problems.* Problems are frequently encountered for which initial or additional training is believed to be a solution. It would appear, then, that this principle is frequently used. However, effective trainers must continually consider whether training should address "nice-to-know" or "need-to-know" issues in the context of the specific training program being planned.

For example, if an identified problem is that it takes too long to serve guests utilizing an operation's drive-thru, then the content of a training program designed to address that issue should focus solely on that specific problem.

✓ ***Training Principle 4:*** *Training Must Emphasize Application.* Most people learn best by doing, and "hands-on" training using an individualized training program is typically the best way to teach many tasks to entry-level employees. However, training can also present information that extends beyond one's position or department. For example, all staff members should learn about their operation's values, vision, and mission. Perhaps some of these issues are addressed during a new employee orientation program (see Chapter 7). However, ongoing formal and informal training opportunities may also be planned for more experienced staff.

✓ ***Training Principle 5:*** *Training Should Consider the Trainees' Current Skills.* The best trainers establish a **benchmark** of what trainees already know and can do and then build on this

Key Term

Benchmark: A standard or point of reference used to measure something else.

foundation of knowledge and skills. This tactic maximizes the worth of training by emphasizing the most important subject matter with which the trainee is unfamiliar. Fortunately, one-on-one training is frequently the training method of choice, and skilled trainers can focus on what the trainee doesn't know instead of repeating what is known.

✓ **Training Principle 6:** *Training Should Be Informal.* The best training is normally personalized and conducted in the workplace with individualized interaction between the trainer and the trainee. It is planned for delivery at a pace that is best for the trainee and that addresses the trainee's specific questions and needs as they arise.

✓ **Training Principle 7:** *A Variety of Training Methods Should Be Used.* Trainees are unlikely to learn when a trainer quickly shows them how to do something but then does not allow the trainees to practice immediately after the training. Most people learn best by doing. High-quality training programs typically include *telling* employees what they must know, *showing* employees what they must know, and then allowing the trainees to *practice* what they must know. Training programs that are created simply to *tell* new workers what to do are rarely effective.

✓ **Training Principle 8:** *Training Should Focus on Trainees.* Good trainers address trainees' needs. They don't try to impress the trainees with their own personal knowledge or skills, nor do they make training more difficult because everyone should "learn it the hard way." Failure to teach a training point "because everyone should know it" is another error, as is the use of jargon or unfamiliar industry terms. Addressing the question, "How would I like to be trained?" often reveals suggestions about tactics that should (and should not) be used.

✓ **Training Principle 9:** *Trainees Must Be Given Time to Practice.* Foodservice staff members in every position must have significant knowledge and skills. Few skills can be learned by reading a book, listening to someone "talk through" a task, or by only watching someone else do it. Rather, skills are typically learned by observing how something is done and then by practicing the activity in a step-by-step sequence. After the task is learned, time and repetition are often needed to enable the trainee to perform the task at the appropriate speed, so adequate practice time is often essential.

✓ **Training Principle 10:** *Trainees Require Time to Learn.* This principle, while seemingly obvious, is sometimes violated. Consider that some foodservice operators expect a new staff member to learn necessary tasks by "tagging along" with an experienced peer. What happens when these trainers must provide a well-thought-out and organized training program even when they are continually interrupted by the work requirements necessary to perform effectively in their own position? Too often it results in poor training results.

✓ **Training Principle 11:** *The Training Environment Must Be Positive.* The stress created when training principle 10 is violated provides an example of a training environment that is *not* positive. As another example, consider someone with training responsibilities who does not enjoy the task. These issues can create a hostile environment without the interpersonal respect necessary for effective training.

✓ **Training Principle 12:** *Trainees Need Encouragement and Positive Feedback.* Most trainees appreciate ongoing input about how the trainer evaluates performance during and after the training is completed.

✓ **Training Principle 13:** *Trainees Should Not Be Forced to Compete Against Each Other.* Contests in which, for example, one trainee "wins" and other trainees "lose" do not encourage teamwork. A better alternative is contests in which all trainees that attain specified performance standards can "win."

✓ **Training Principle 14:** *Teach Only the Correct Way to Perform a Task.* Showing a trainee how something should *not* be done does little good, but this happens when, for example, a trainer notes "Here's the wrong way that many employees use . . . ." Instead, use the correct work methods on a step-by-step basis, with the trainer's presentation followed by the trainee's demonstration.

✓ **Training Principle 15:** *Train One Task at a Time.* Tasks should be taught separately, and each should be broken into steps that are taught in proper sequence.

✓ **Training Principle 16:** *Train Each Task Using a Step-by-Step Plan.* Consider the task of taking a guest's food order. Beginning with the task's first step, a trainer should present correct procedures and encourage the trainee to demonstrate them. Trainer feedback helps the trainee to identify where performance improvements could be helpful. After the trainee demonstrates the step, repeat the process until all steps are presented and successfully demonstrated by the trainee. The trainer can then demonstrate the correct procedures to do the entire task again, and the trainee can then repeat the correct procedures and practice each step to build the appropriate speed for acceptable task performance.

✓ **Training Principle 17:** *Trainees Should Know Training Requirements.* Experienced trainers often use a "preview, present, and review" sequence. They tell the trainees what they are going to say (preview), they tell them (present), and they tell them once again (review).

✓ **Training Principle 18:** *Consider the Trainees' Attention Span.* Several short training sessions are generally better than one long session. Consider the complete range of subject matter to be presented, then break the total training requirement down into manageable (short) parts for each training session.

✓ **Training Principle 19:** *Learning Should Be Paced.* Learning that is spread out to address one or just a few concepts at a time allows trainees to practice and improve on basic skills in a focused way. They can concentrate on one or several skills rather than on all skills and better learn the correct way to perform all of them.

✓ ***Training Principle 20:*** *Learning Speed Varies for Trainees.* Individualized training allows the trainer to incorporate only what the trainee doesn't know, and exclude what the trainee does know, from the training process. As a result, the training time required may vary from one trainee to the next.

Find Out More

Those who have experience in training employees recognize that they often encounter generational differences in the learning styles of their trainees. While often loosely defined, a generation is simply a group of people born during a particular time. Very often, individuals in these generations operate with similar attitudes, ideas, and values, some of which can affect their learning styles.

While examples of generational differences abound, one easy way to recognize situations that can directly affect employee training involves familiarity with the use of technology. For example, the "Baby Boomer" generation (those born between 1946 and 1964) currently make up approximately 25% of the workforce. Millennials (Generation Y) (those born between 1981 and 2000) currently make up approximately 35% of the workforce. In most cases, the exposure of the Baby Boomer generation to advanced technology is often less useful. Most members of this generation still recall when answering the phone meant going to a physical location and picking up a receiver.

Millennials, however, grew up with advanced technology and are quite comfortable with the use of it in many settings. It is easy to see that the use of technology may be one area where trainers need to consider generational differences as they plan the delivery of their training programs.

To learn more about some other areas in which generational differences can impact training, enter "generational differences in the workplace" in your favorite search engine and review the results.

What Would You Do? 8.1

"Seriously, you want me to train a new server again?" said June Vermonte, a long-time server at the Pink Flamingo restaurant.

"You're our best server, and you know it, June. I just want to make sure that our new hire gets the best possible chance of success, and he will if you do the training," replied Demonte Jackson, the dining room supervisor at the Pink Flamingo.

Demonte had just hired a new server to replace one that had recently resigned from the restaurant. They were discussing the newly hired server's initial training, and Demonte wanted June to assume that responsibility.

"Here's the thing Demonte," said June, "when I train a new server, I get less tables because of the time it takes to show them the ropes. Less tables means less tips for me. And then, to be fair, I always feel like I need to give at least some of the tips I do get to the new trainee because they do assist a little bit. The bottom line is, every time you want me to be the trainer, I end up paying for it! That's not fair!"

Assume you were Demonte. Do you agree that the training plan for the new server is unfair to June? What could you do to ensure that the new employee gets the best training possible, but in a way that would make June happy to serve as the new employee's trainer?

Characteristics of Effective Trainers

Every foodservice operator will, from time to time, experience the need to train new team members. In some cases, foodservice operators may be able to provide the training themselves, but in other cases they will not. This is especially so when a foodservice operation is large, and then other individuals within the operation must be chosen to provide new employee training.

Who should provide the training? Sometimes this question is answered by asking other questions: Who is available? Who wants to do it? Who has the time? Who will complain least if given the assignment? Who is a good "people person" who can interact well with the new staff member?

While these factors are often relevant, others are more important, including the following characteristics which are important ones for effective trainers:

✓ *Have the desire and resources to train*—In most cases, effective trainers want to train for reasons that include helping others, internal recognition for a job well done, and the knowledge that effective trainers are frequently promoted to higher-level positions.

Unfortunately, in some cases, there are also reasons why training might not be an attractive assignment. This includes the expectation that the trainer will complete all their regularly assigned tasks and still conduct the training. Also, a trainer might want to do a good job but cannot because they have not been taught how to train and/or because there is insufficient time, equipment, money, or other resources required to do so. Experienced foodservice operators know that stress resulting from inadequate training resources is often a disincentive for accepting and successfully completing a training assignment.

✓ *Have the proper attitude about the employer, peers, position, and the training assignment*—Foodservice operators who emphasize the importance of staff members and provide quality training opportunities to employees at all levels increase the morale of their trainers. Conversely, when training is just another and not so important responsibility, a less-than-willing attitude toward being a trainer can result.

✓ *Possess the necessary knowledge and skills to do the job for which training is needed*—Effective trainers must be knowledgeable about and have the skills necessary to perform the work tasks for which they will train others.

✓ *Use effective communication skills*—Trainers are effective communicators when they (1) speak in a language that the trainee understands, (2) recognize that body language is a powerful method of communication, (3) use a questioning process to learn what the trainee has learned, and (4) speak to effectively communicate rather than to impress.

✓ *Know how to train*—The importance of train-the-trainer programs to give trainers the necessary skills to train effectively should be obvious but often is overlooked.

✓ *Have patience*—Few trainees learn everything they must know or be able to do during their first exposure to training. Effective trainers understand that training must sometimes be repeated several times and in different ways. They know the goal is not to complete training quickly; rather, it is to provide the knowledge and skills the trainee needs to be successful.

✓ *Have time to train*—Effective training takes time, and it must be scheduled for the trainer and for the trainees.

✓ *Show genuine respect for the trainees*—Trainees must be treated as professionals. Trainers know that those whom they respect will also respect them, and mutual respect allows training to be more effective.

✓ *Are enthusiastic*—Newly employed staff members want reinforcement that their decision to join a foodservice operation was a good one. Initial experiences with an enthusiastic trainer help to develop the foundation for successful training and for team members' long-term commitments. Trainers can reinforce the philosophy of more senior staff: "This is a good place to work; let's make it a better place to work, and this training will help us do that."

✓ *Celebrate the trainees' success*— Experienced trainers are familiar with the saying that, "If a trainee hasn't learned, it is because the trainer hasn't trained!" A successful trainer is one who has successfully trained others. The reverse is also true: trainers have not been successful when their trainees have not learned. The best trainers take time to celebrate when learning occurs.

✓ *Value diversity*—Most foodservice operations employ workers with a variety of backgrounds and cultures and are strengthened because of the different perspectives

Find Out More

For those employees in a foodservice operation who will be assigned training responsibilities, the completion of a formal train-the-trainer program is always a good idea. Train-the-trainer programs are designed to equip trainers with the skills needed to deliver high-quality training programs in ways that have the most positive impact on trainees.

Those assigned to training responsibilities must be well-respected, experienced, patient, and passionate. Trainers will also need to have strong leadership skills, communication capabilities, and exceptional listening skills. When those personal characteristics are added to the teaching skills that can be learned through the successful completion of a train-the-trainer program, trainers, trainees, and the foodservice operation itself will benefit.

Train-the-trainer programs are available in a variety of lengths and costs. To review the content of several such programs, and to see which ones might most benefit your own operation, enter "foodservice train-the-trainer programs" in your favorite search engine and review the results.

these individuals can provide. An effective trainer accepts the challenge of developing all trainees to the fullest extent possible, even though training tactics might differ based on the trainees' cultural or demographic backgrounds. For example, group trainers may need to actively solicit question responses from trainees who don't readily participate in discussions, and train ees from some cultures may be embarrassed to participate in **role-playing exercises.**

Key Term

Role playing (exercise): An activity that allows trainees to act out a specific situation or assume the role of another person in a real-to-life situation or scenario.

Steps in Creating Effective Training Programs

The creation of effective training programs is an on-going activity. As shown in Figure 8.1, to plan an effective training program operators must go through a specific five-step process that requires them to:

Step 1: Identify Training Needs
Step 2: Develop Training Objectives
Step 3: Develop Training Plans

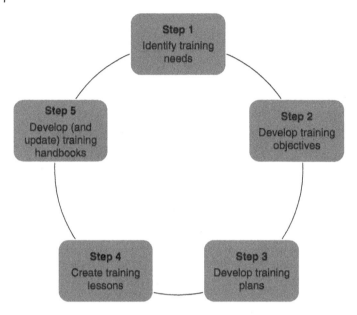

Figure 8.1 Steps in Creating Effective Training Programs

Step 4: Create Training Lessons
Step 5: Develop (and Update) Training Handbooks

Each step in the planning process is important and properly understanding and completing the step is essential if a foodservice operation is to develop high quality training programs.

Step 1: Identify Training Needs

Every foodservice operation has training needs, but the specific needs of an operation vary depending on the operation itself. There are numerous examples of short-term operating and other concerns that can be addressed by training. New work methods, the purchase of new equipment or technology, and/or implementation of cost-reduction processes may be necessary to enable the hospitality operation to better meet operational goals. Also, there may be a need for more or different products and services to meet guests' evolving needs.

If only the "squeaky wheel gets the grease," foodservice operators would have little difficulty in knowing about problems that require priority resolution with staff training. However, there are also "silent" problems that can create difficulties, and many of these can also be addressed with training.

Regardless of the topics to be addressed, all training programs must be **cost-effective**.

To be cost-effective a training program must provide time and money benefits that outweigh its costs. To do so, training must also be **performance based**. It should be planned and delivered in a systematic way to help trainees become better able to perform the tasks that are essential for their jobs. The success of training can then be demonstrated by considering the extent to which knowledge, skills, and performance improve because of the training.

Operators can determine their own training needs in several ways.

Key Term

Cost-effective (training): A term that indicates that an item or activity such as training is worth more than it costs to provide it.

Key Term

Performance-based (training): A systematic way of organizing training to help trainees learn the tasks that are essential for effective on-job performance.

✓ *Listening to guests.* Training programs have been developed to enable guests to receive what they want, and an objective drives the training to perform this task. For example, there may be a preferred way to make a dinner reservation for a guest at 8:00 PM. Applicable employees should know how to best ensure that a table for the guest will be available at that time (not 30 minutes later!). The tactics involved in the training for "taking reservations" is the method that should be taught to new service employees who might, as part of the job, "take reservations."

As a second example, a person in line for a food truck meal orders French fries "without salt." Since this is a common request by guests, the cook that prepares the fries has been taught the correct way to prepare this salt-free menu item.

In both the above examples, affected guests might inform an employee they recognize that there were changes in the standards (the guest did not have a table at a desired time, and/or fries were served with additional salt). Depending on how frequently these two problems occurred, training and/or re-training needs could be identified, and training could be undertaken to best ensure training needs meet requirements.

✓ *Listening to current staff members*—Some foodservice operators use suggestion boxes, open-door policies, and frank input from performance appraisals and coaching sessions to identify problems that can be resolved by training.

Current staff can be an excellent source of information about necessary training, and it can take two forms. The first is that employees can reflect on their own training and use hindsight to help foodservice operators better understand where initial training programs could be improved. Secondly, they can consider their current positions and identify areas where their performance could be improved with additional training.

✓ *Observing current work performance*—Those who "manage by walking around" may note work procedures that deviate from established standard operating procedures (SOPs). It is important that proper performance is clearly defined so employees know what is expected of them, and foodservice operators know when performance is acceptable.

A training goal is to teach a trainee how to correctly perform a task and the definition of *correct* refers to both quality and quantity dimensions. Task performance standards should be reasonable (challenging but achievable). Performance standards must also be specific so they can be measured. When they are, it will be possible to identify whether training in a specific task has been effective.

✓ *Reviewing inspection scores*—All foodservice operations are subject to inspections by local health departments. In addition, foodservice operators themselves may create internal inspections for safety-related conditions and proper opening or closing procedures. Franchised operations are regularly inspected by their franchisors. When inspection scores are carefully reviewed, it can help identify areas where additional training may improve the scores.

✓ *Analyzing financial data*—Differences between budget plans and actual operating data may suggest negative variances traceable to problems with training implications. After problems are identified, corrective actions that include training may be implemented.

✓ *Conducting **exit interviews**—Formal or informal discussions with resigning employees may identify training topics to help reduce future turnover rates and improve operations.

Exit interviews can be of great value because departing employees are likely to be more honest in their assessments of what their job experience was like than those who are still in their jobs. In many cases, insight into recruiting, on-boarding, and training needs may be revealed in an exit interview.

Key Term

Exit interview: A formal meeting between a representative of a foodservice operation and a departing employee.

The methods in which foodservice operators identify their training needs are varied, and as many different methods as possible should be utilized.

Step 2: Develop Training Objectives

Training objectives specify what trainees should know and be able to do when they have successfully completed their training. Those who plan training programs must know what the training should accomplish, and training objectives help them do this.

Training objectives are critical to training evalua-tion, and they should describe the expected results of the training rather than the training process itself. For example, assume a foodservice operator has identified the need to train servers in the proper use of the operation's **point-of-sale (POS)** system. In this situation, the proper training objective is to ensure all trainees can properly use the POS sys-tem, and this goes beyond a point where the trainee has completed a training lesson to teach them how to use it.

Key Term

Point-of-sale (POS) system: An electronic system that records foodservice customer purchases and payments, as well as other operational data.

One good way to develop training objectives is to complete the following sen-tence for each training need identified in Step 1:

> *As a result of the satisfactory completion of the training session, the trainee will . . .*

Training objectives must be reasonable (attainable), and they must be measur-able. Objectives are not reasonable when they are too difficult or too easy to attain. The concern that training objectives be measurable relates to the role of the objec-tives in the training evaluation process. For example, how can training programs be evaluated if success is measured by objectives such as those listed below?

✓ Trainees will *realize* the importance of proper POS system usage.
✓ Trainees will *understand* how the POS system works.
✓ Trainees will *recognize* the need to use the POS system properly.

Contrast the above with measurable objectives for the same topics:

✓ Trainees will *demonstrate* they can log their own server identification informa-tion into the POS system.
✓ Trainees will *show* their trainer that they can properly enter guests' orders into the POS system.
✓ Trainees will *confirm* to their trainer that they can properly produce a guest check total suitable for presentation at the end of a guest's meal.

The best training objectives typically use an action verb to tell what the trainee must demonstrate or apply after training. Examples include *operate, calculate, explain, show, produce,* and *assemble.* By contrast, unacceptable verbs that cannot be readily measured include *know, appreciate, believe,* and *understand.*

Training objectives are developed used for two main purposes:

1) To help the trainer connect the purpose(s) of the training program with its content. Specific reasons for training become clear when training needs are

defined (Step 1), and when the measurable results of training are clearly indicated.

2) To help evaluate the impact of training (see Chapter 9).

Step 3: Develop Training Plans

Training plans organize training content and provide an overview of the structure and sequence of the entire training program. They show how individual training lessons should be sequenced to teach required knowledge and skills.

Key Term

Training plan: A description of the overview and sequence of a complete training program.

Suggestions for determining the sequence for subject matter in a training plan include:

✓ Beginning with an introduction explaining why the training is important and how trainees will benefit from it.

✓ Providing an overview of training content.

✓ Planning training lessons to progress from simple to complex. Simple information at the beginning allows trainees to quickly feel comfortable in the learning situation. It also provides the confidence needed to master the program.

✓ Building on the trainees' experiences. Combine unfamiliar information with familiar content to allow trainees to build on their experience.

✓ Presenting basic information before more detailed concepts are discussed.

✓ Progressing from general information to specific information. For example, provide an overview of how to set up a dining room and then begin with specific steps.

✓ Considering the necessity of "nice-to-know" and "need-to-know" information. Basics should be presented before other information and addressing "why" before the "how" is generally best.

✓ Using a logical order and clearly identify what information is prerequisite to other information.

Training plans allow trainers to (1) plan the dates and times for each training lesson, (2) consider the topic (lesson number and subject), (3) state the training location, (4) indicate the trainer('s) who are responsible for conducting the training, and (5) determine the trainees for whom specific training lessons are applicable.

To illustrate, assume a trainer is planning a training program for a new employee who must learn how to correctly perform all tasks in a position. Perhaps some tasks can be taught in one training session. By contrast, several (or more) training sessions may be needed to teach just one other task. The training plan allows planners to think about what must be taught (training lessons) and the sequence and duration of each training lesson.

When a training plan is developed to teach all tasks to a new staff member, training dates and times must only consider the trainer's and employee's availability. If the training impacts several employees, other tactics may be necessary. Perhaps each session will be planned for alternate dates and times to accommodate all affected personnel. The training location could be the same for every session, or it could accommodate group training in a congregate setting and individual training at specific workstations. The trainees might include all staff members for some sessions and only selected staff members for other sessions.

Technology at Work

Some foodservice operators who are responsible for significant amounts of training may want to join the Council of Hotel and Restaurant Trainers (CHART).

CHART is a non-profit association of hospitality training and development and human resource professionals representing trainers in many segments of the hospitality industry including quick service, fast casual, casual, full service, fine dining, hotels, resorts, clubs, colleges and universities, casinos, cruise ships, senior living, and convenience stores.

CHART membership is open to individuals who have the responsibility to create, implement or oversee programs for education, training, and developing human resources.

In addition to holding annual conferences, CHART publishes newsletters and creates web postings that help trainers better learn their roles and become more effective in their jobs.

To learn more about CHART, enter, "Council of Hotel and Restaurant Trainers" in your favorite search engine and view the results.

Step 4: Create Training Lessons

A **training lesson** provides all information needed to present a single session that is part of a broader training plan. In effect, it is a "turn-key" module that defines a specific training session:

✓ *Why*—the objective(s) of the training session
✓ *What*—the content of the training session
✓ *How*—the method(s) used to present the training

Key Term

Training lesson: Information about a single session of a training plan. It contains one or more training objectives and indicates the content and method(s) to enable trainees to master the content.

Individual training lessons may be as short as a few minutes or as long as several hours. A training lesson may be needed to teach new staff members how to perform a single task such as operating an oven, or it can be used to teach experienced staff new steps in a single task.

Trainers can use numerous resources to develop training content, including:

✓ Manufacturers' operating manuals for equipment
✓ Standard operating procedure (SOP) manuals for their own operations
✓ Task breakdowns for positions
✓ Applicable books/magazines and internet sources
✓ Industry best practices
✓ Training resources from professional associations such as the National Restaurant Association (NRA) and the American Hotel & Lodging Association Educational Institute (AHLEI)
✓ Materials available from vendors
✓ Ideas from other hospitality organizations
✓ Trainers' notes from other training sessions
✓ The trainer's own experience

Training lessons may be designed to be delivered to one trainee at a time or to groups of trainees (see Chapter 9).

Step 5: Develop (and Update) Training Handbooks

Designing effective training programs requires thought, time, and creativity, and the process is most cost-effective when the training plans, lessons, and resource materials developed are used for more than one training experience. For that reason, experienced foodservice operators know they must develop (and keep updated) an easy-to-utilize **training handbook** specific to their own operations.

Key Term

Training handbook: A hard copy or electronic manual containing the training plan and associated training lessons for a foodservice operation's complete training program.

A wise foodservice operator will maintain their training information in an organized fashion that allows easy replication of training, with revision to the handbook, when necessary. This benefits operators because the time and money spent to develop training tools need not be replicated.

It is important to recognize that a training handbook is a tool to be used by a trainer, rather than a self-training guide for a new employee. While it certainly may be the case that some amount of reading material may be required of new employees, the employee handbook is a guide for trainers to ensure they address important topics that must be known by those in each position in their operation.

A handbook used in training a new staff for the required tasks and information to be learned in their positions may include:

✓ An introduction with instructions for handbook use.
✓ A listing of concepts and training lessons that must be completed for all new employees. Examples of important concepts or key information which must be known by all employees may include topics such as fire safety, the procedures to be used to report harassment allegations, and information about where to find the trainee's weekly work schedules.
✓ A listing of concepts and training lessons that must be completed for trainees holding a specific position. For example, training lessons addressing food safety and the importance of properly handling **time and temperature control for safety foods (TCS food)** will be essential for all foodservice production workers, but an operator may decide that such information is not useful in the training of hosts or servers.

> **Key Term**
>
> **Time and temperature control for safety food (TCS food):** Foods that must be kept at a particular temperature to minimize the growth of food poisoning bacteria or to stop the formation of harmful toxins.

Similarly, detailed instruction on how to enter guest orders in the operation's POS system will be important for all servers who are taking guest orders and processing guest payments. However, this information will not be available for food production workers who have no role in entering data into the POS system.

✓ The actual position specific training lessons to be administered to applicable employees.
✓ The method(s) to be used to assess trainee learning. In some cases, this may consist of a test or a demonstration of a learned skill.
✓ The method(s) to be used to document training delivery. This is a critical part of the training handbook because it is often essential to document that training has occurred.

For example, most foodservice operators will want to document the fact that training regarding sexual and other types of harassment was provided. This is done to help minimize the possibility of facing legal consequences that can be avoided when the training was documented. The training manual gives trainers the information they need to ensure that proper documentation has been prepared. This documentation may then be placed in the trainee's personal file or in a separate file designated for documenting training program delivery (see Chapter 9).

Technology and User-Generated Content (UGC) Scores

Advancements in technology have radically changed almost every phase of successful foodservice operations and this includes training. Consider, for example, social media and delivery operations and point-of-sale (POS) systems for accounting purposes. Successful foodservice operators must learn about their guests' needs and the extent to which these needs are met. Surveys can help identify training challenges, and on-going in-person interactions with guests can also be helpful to identify training needs. Today, however, the most important input from guests is received from a careful and continual review of popular **user-generated content (UGC) sites**. These are important to marketing and ultimately to profitability, and foodservice operators should pay special attention to UGC sites that can identify problem areas in food production or service and thus influence training-related decisions.

Key Term

User-generated content (UGC) site: A website in which content that includes images, videos, text, and audio have been posted on-line by the site's visitors. Examples of currently popular UGCs include Instagram, X (formerly Twitter), Threads, Facebook, Pinterest, and YouTube.

Many UGC sites enable the site's users to post reviews of foodservice operations they have visited. Also, some UGC sites significantly feature restaurant reviews or are devoted almost exclusively to these types of reviews. These reviews may be the best source of information for foodservice operators who want to identify possible "problem" areas in their operations.

While the popularity of any given UGC review site will likely increase or decrease over time, Figure 8.2 lists 10 of the most popular UGC review sites currently in operation. Third-party delivery company sites such as Grubhub, DoorDash, and Uber Eats also publish user reviews.

Foodservice professionals reviewing best-selling books published in the 2010s and earlier will find virtually no information about the importance of obtaining guest feedback on UGC sites displaying restaurant reviews. However, times change, and today's UGC sites featuring foodservice operating reviews are

Yelp	Google My Business
OpenTable	Facebook
TripAdvisor	Foursquare
Zagat	Gayot
Zamato (formerly Urbanspoon)	The Infatuation

Figure 8.2 Popular UGC Review Sites

very popular. Their impact on identifying needed areas of training is well established and highly significant.

Published customer reviews of restaurants are not new. Beginning in the late 1800s, the New York Times regularly published restaurant reviews written by food critics employed by the newspaper, and that newspaper still does so.

In 1979, Tim and Nina Zagat began the "Zagat Survey." The philosophy behind the Zagat Survey was unique in that, instead of reading one review written by a paid restaurant critic, the public would be better served by a rating based on hundreds of guest responses. By compiling data from surveys of actual customers, the Zagat Survey rated restaurants on a 30-point scale using the categories of food, décor, service level, and cost. In 2016, Zagat (then owned by Google) switched to a system that scores restaurants on a scale of one to five stars. However, the idea of actual users (not only professional reviewers) sharing their dining experiences proved to be extremely popular. Today a variation of this Zagat approach is still used, and many popular UGC sites allow users to post their restaurant review comments directly on the site.

The importance to foodservice operators of listening to their guests by monitoring UGC sites featuring restaurant reviews was demonstrated in a 2022 survey that found:

✓ 99% of consumers have used the Internet to find information about a local business in the last year.

✓ More consumers are reading online reviews than ever before. In 2021, 77% "always" or "regularly" read them when browsing local businesses including foodservices (up from 60% in 2020).

✓ The percentage of people "never" reading reviews when browsing local businesses has fallen from 13% in 2020 to just 2% in 2021.

✓ 89% of consumers are "highly" or "fairly" likely to use a business that responds to online reviews.

✓ 57% say they would be "not very" or "not at all" likely to use a business that doesn't respond to reviews.

✓ In 2021, just 3% of respondents said they would consider using a business with an average star rating of two or fewer stars. Note: this is down from 14% in 2020.[1]

UGC sites display the rankings of foodservice operations that have been reviewed. Essentially, the "score" given on a UGC site featuring guest reviews allows the reviewer to rank the quality of the experience that occurred when the reviewer interacted with the operation. When operations have met or exceeded their expectations, they are highly ranked, and lower rankings are given when the reviewer's expectations were not met.

---

1 https://www.brightlocal.com/research/local-consumer-review-survey. Retrieved August 30, 2023.

Various UGC sites featuring user reviews have different scoring systems. For example, on Yelp, scores can range between 1 and 4 "Stars." On Google and Facebook, the scores range between 1 and 5 "Stars." On TripAdvisor reviewers assign a score of 1–5 that results in the display of filled or partially filled "Circles."

Grubhub uses a 5 "star" rating system. It also publishes an additional summary, "Here's what people are saying," about a business. For example, the review for a single operation might say, "Here's what people are saying: '93% Food was good, 88% Delivery was on time, 82% Order was correct.'"

On Google and TripAdvisor, among others, the sites' operators also use algorithms to rank operations. For example, a search for "Best pizza near me" will produce a **search engine results page (SERP)** that ranks pizza restaurants "in score order."

A search engine results page (SERP) lists results that a search engine returns in response to a user's specific word or phrase query. The specific scoring and ranking systems used by UGC review site operators vary. However, an essential fact is that foodservice operators must recognize that the scores matter to guests.

> **Key Term**
>
> **Search engine results page (SERP):** The web page of a search engine such as Google (used by Apple), Bing, or Yahoo shows a user when a reader types in a search query.

Online market research indicates that 70% or more of site visitors reviewing online ratings will only continue to read information about a specific operation if its ratings are within the top two scores (3 and 4 for a four-point ratings system, and 4 and 5 for a five-point rating system). Therefore, scores achieved by an operation directly affect the amount of business they will do. In fact, a comprehensive study at the Harvard Business School found that a 0.5-star increase in average Yelp rating generates a 9% revenue increase for an independent restaurant.[2]

Foodservice operators monitoring their on-line reviews can learn what their customers think about them. This knowledge can lead directly to areas within an operation where additional training could yield increased UGC site scores. While some foodservice operators still conduct formal on-paper or on-line guest surveys to learn what their guests think about them, in most cases, guest reviews posted on-line provide the same information, and it is current and on-going.

Effective foodservice operators know that excellence in training program development requires effort. When training programs have been carefully planned and prepared, however, it is easier to implement the training. The critical nature of implementing, presenting, and evaluating team member training, and how to do those things well, is so important to the long-term success of a foodservice operation that it is the sole topic of the next chapter.

---

2 https://dl.acm.org/doi/pdf/10.1145/3432953. Retrieved August 30, 2023.

What Would You Do? 8.2

"The vendor promised the new system would make things faster and better. Well, I think we can both agree the new system has made things slower and worse!" said Marsha, the manager of the Casual Café.

Marsha was talking to Sharon Cronin, the café's owner. Sharon was concerned because many guest reviews recently posted online about the Casual Cafe complained about mixed-up orders and slow service.

These types of reviews had only come recently, following Sharon's decision to purchase and install the "At The Table" point-of-sale (POS) system. At The Table was a new program that allowed servers to take guest's orders with a handheld order entry device. Guest payments were also processed with the handheld device. This, in turn, reduced the number of times a server had to walk back and forth from a guest's table to a central POS data entry terminal.

Since much of the Casual Café's business was done in its outside dining area, the trips to the POS terminal took a lot of time. Sharon had agreed with the product vendor that reducing the time servers walked to and from the outdoor area along with increasing the amount of time they could interact with guests, would be a good thing.

The problem, unfortunately, was that several servers had difficulty mastering the new technology. It now took more time to enter orders in the new system than it had in the old system, and the error rate for order entry was also higher than previously.

"I agree we have some problems," said Sharon "but I'm not sure the problems have to do with the program as much as they have to do with us!"

Assume you were Sharon. Do you Agree with Marsha that the "program" is likely the cause of the increased customer complaints, or do you think that ineffective staff training is more likely the problem? What would you tell Marsha to do about it?

Key Terms

Training
Train-the-trainer
  (program)
Off-the-shelf (training
  programs)
Warm body syndrome
Benchmark
Role playing (exercise)
Cost-effective (training)

Performance-based
  (training)
Exit interview
Point-of-sale
  (POS) system
Training plan
Training lesson
Training handbook

Time and temperature
  control for safety food
  (TCS food)
User-generated content
  (UGC) site
Search engine results
  page (SERP)

Operator's 10-Point Tactics for Success Checklist

Evaluate your need for, and the current status of, each of the following operational tactics. For those tactics you think are important, but not yet in place, develop an action plan for its implementation including who will be responsible for the tactic's completion and the target date by which it should be completed.

Tactic	Don't Agree (Not Done)	Agree (Done)	Agree (Not Done)	If Not Done Who Is Responsible?	Target Completion Date
1) Operator recognizes the benefits that accrue from implementing effective training programs.	____	____	____		
2) Operator can identify obstacles to, and myths associated with, the development of training programs.	____	____	____		
3) Operator can identify principles of training that impact training program effectiveness.	____	____	____		
4) Operator can identify the personal characteristics that will be possessed by effective trainers.	____	____	____		
5) Operator recognizes that the identification of training needs is the first step in the planning of employee training.	____	____	____		
6) Operator recognizes that the development of training objectives is an important step in the creation of effective training programs.	____	____	____		
7) Operator recognizes that the development of training plans is an important step in the creation of effective training programs.	____	____	____		

	If Not Done				
Tactic	Don't Agree (Not Done)	Agree (Done)	Agree (Not Done)	Who Is Responsible?	Target Completion Date
8) Operator recognizes that the creation of training lessons is an important step in the creation of effective training programs.	___	___	___		
9) Operator can state the purpose of developing and updating training handbooks.	___	___	___		
10) Operator recognizes the impact of technology and UGC site scores on the development of training programs.	___	___	___		

9

Delivering and Evaluating Training

What You Will Learn

1) The Importance of On-Job Training
2) The Four Steps in On-Job Training
3) Group Training Methods
4) How to Evaluate Training Efforts

Operator's Brief

The previous chapter addressed how foodservice operators plan effective training programs. Carefully planned training, however, will be ineffective if it is not delivered correctly. While the implementation of staff training can take a variety of forms, one good way to examine training delivery is by classifying it as either individual training or group training.

In this chapter, you will learn that the most popular form of individual training in the foodservice industry is on-job training, and it is a four-step process:

✓ Step 1: Preparation
✓ Step 2: Presentation
✓ Step 3: Practice and demonstration
✓ Step 4: Coaching

Trainers first prepare trainees by identifying training objectives and explaining the importance of the training. Presentation of training requires that tasks be divided into separate and teachable steps. Trainers explain the first and all subsequent steps in the task and answer trainee's questions. After the presentation of training information, allowing trainees to practice and demonstrate they have mastered a task is especially important, and you will learn why that is so. Coaching is the final activity in on-job training, and it allows trainers to follow-up on training and ensures lessons have been "taught" and have actually been put into practice.

In some cases, group training is the most effective way to deliver information. In this chapter, you will learn how to prepare for group training and how to best deliver it.

Whether you choose to use individual on-job training, group training, or a combination of both, there are several special training-related considerations to be recognized and the most important of these are addressed in this chapter.

Regardless of the method used to deliver training, evaluation of training efforts is essential. Training evaluation that occurs before, during, and after training is critical to ensure that expenditure of training resources represents the best use of these funds.

Revisions to training identified by evaluation will allow you to continue your journey of operational improvement. In this chapter, you will learn how you can assess training results to ensure its impact is optimized and to make any needed improvements.

CHAPTER OUTLINE

On-Job Training
On-Job Training Steps
    Step 1: Preparation
    Step 2: Presentation
    Step 3: Practice and Demonstration
    Step 4: Coaching
Other Individual Training Methods
Group Training
    Preparing for Group Training
    Delivering Group Training
Special Training Considerations
Evaluating Training Efforts
    Evaluation Before Training
    Evaluation During Training
    Evaluation After Training

On-Job Training

After training efforts have been planned, foodservice operators can choose from a variety of training program delivery methods. When **on-job training** is utilized as a training method, the trainer teaches job skills and knowledge to

Key Term

On-job training (method): An individualized training method in which a knowledgeable and skilled trainer teaches a less experienced staff member how to properly perform job tasks.

one trainee at a time; typically at the trainee's assigned workstation. In many ways, it is the best training method because an individual trainee can learn at their own pace, and immediate feedback can be provided.

While it is effective, in some cases on-job training is not done properly. As suggested by other common names (for example, "tag-along" training and "shadow" training), some foodservice operators incorrectly believe trainees can learn simply by watching and helping a more experienced peer. Unfortunately, this approach ignores several key training principles (see Chapter 8). Also, trainers who have not learned how to properly train, who must do so in addition to completing their own existing full-time responsibilities, and who have been taught by equally unskilled trainers are not likely to be effective on-job trainers.

One primary purpose of on-job training is to provide necessary knowledge and skills to new employees. Other purposes can be to teach current employees how to perform an existing task in a different way and to learn how to perform new tasks.

Individual on-job training can be supplemented by group training that can be effective when several team members must learn the same thing. For example, assume all cooks must learn how to operate a new piece of equipment. Demonstration is likely to be an important tactic in this training, because the process can be very effective with a small group of cooks instead of just one cook at a time.

Some foodservice operators prefer on-job training for the wrong reasons. They think it is easy, fast, and inexpensive because, "*All you must do is allow a trainee to follow an experienced staff member who can teach the trainee what is needed to do the job.*"

This perception of on-job training is incorrect. First, the needs to define training requirements and develop training objectives, training plans, and training lessons addressed in the previous chapter are important for on-job training. Also, trainees must be prepared for the training, and this involves more than just saying, "Tag-along with JoAnn, and she'll show you what to do."

On-job training can be simple to implement because the steps involved in its planning and delivery are not complicated. The process takes time, however, and a commitment of financial resources is required to effectively plan and deliver it. On-job training is most effective as a training method when:

✓ *It incorporates basic training principles*—Training should focus on those being trained. Trainers should consider the trainees' backgrounds, training should be organized and, when possible, informal. Trainees should be allowed to practice, their individual attention spans and learning speeds should be considered, and appropriate training tactics are needed. For example, one trainee may like to ask questions, another may want to practice before moving to the next step, and still another trainee might attain performance standards without repetitive explanations or demonstrations.

✓ *It provides maximum realism*—Training must focus on real workplace problems, and the training should help provide solutions to these problems.

✓ *It provides immediate feedback*—Effective on-job training allows a trainee to demonstrate (practice) what has been learned as soon as the training is completed. The trainer can observe the trial performance and recognize proper or address improper performance. In both instances, correct performance can be encouraged.

✓ *It can be used to train new and experienced staff*—On-job training can teach new staff members the tasks they must perform, and it can also teach experienced staff about revised work methods for one or more new tasks.

✓ *It is delivered by peers who regularly perform the task*—The trainer can be a role model, teamwork can be encouraged, and a corporate culture that encourages cooperation and mutual assistance can be fostered.

When undertaken properly there are very few disadvantages to on-job training. If implemented improperly by foodservice operators, however, there can be several potential disadvantages:

✓ Experienced staff members who have not learned how to train can make training-related errors that prevent trainees from optimizing learning.

✓ Effective on-job training requires a step-by-step approach, and, if the steps are not followed in order, training results may be poor.

✓ It could ignore correct work procedures if those used by the trainer vary from the standard operating procedures (SOPs) established by the operation.

✓ It can lower the morale of a work team when experienced employees who can do their jobs correctly but who do not know how to train become frustrated with their training responsibilities. This problem can be compounded when trainers must complete their own work responsibilities while still performing on-job training tasks.

Improperly planned and delivered on-job training is not cost-effective because the desired result (an employee who can perform tasks meeting performance standards) is not likely to be attained. It is even less effective if the training results in frustrated trainers and trainees who have not received the organizational support that is required. This in turn can result in high employee turnover rates and/or unhappy employees (and guests!). As a result, foodservice operators who get excellent on-job training results generally follow a four step on-job training process.

On-Job Training Steps

The effective use of on-job training requires the use of four distinct steps, as shown in Figure 9.1.

As shown in Figure 9.1, the first step in on-job training involves various preparation activities and the second step is the actual presentation of the training. The third step allows practice and demonstration by the trainee and a final step involves various **coaching** and related activities that are the trainer's responsibility.

Key Term

Coaching: A training and supervisory activity that involves informal on-the-job conversations and/or demonstrations that encourage proper (and discourage improper) work behavior.

Step 1: Preparation

Proper preparation is the first step in effective on-job training. Several principles are useful when preparing for on-job training and each will have been considered if the planning for training steps noted in Chapter 8 were followed:

✓ *State training objectives*—Training objectives for the entire training program must be available in the training plan and for each segment of the training when specific training lessons are discussed. For example, assume a training lesson addresses how to prepare one baked dessert (and there are several desserts on the menu). In this example, the trainer will know that the training objective is to "prepare baked fruit pies following their applicable **standardized recipes**."

Key Term

Standardized recipe: The instructions needed to consistently prepare a specified quantity of food or drink at the expected quality level.

✓ *Review the specific tasks to be taught*—Before beginning any training, the trainer who will teach an employee how to prepare a fruit pie should review the applicable steps in the recipe. The reason: These steps will be the most important parts of the training content.

✓ *Consider the training schedule*—The training plan should indicate the length of the training activity and where in the overall training for dessert preparation activities should occur. Experience may suggest that training for one fruit pie recipe normally takes 25 minutes. Each recipe can be taught at any time after workplace safety and equipment operation has been taught, so the trainer has flexibility. The best time to conduct training is when production requirements are low, when employees who normally work in the area are working elsewhere or on a break, and when the trainer has adequate time to provide the training.

✓ *Select training location*—When practical, training should occur in the actual workstation where the task will be performed. In this example, training in dessert preparation is planned for the operation's bake shop area.

✓ *Assemble training materials/equipment*—The training lesson and supporting standard recipe should be in the training handbook, which should indicate all necessary materials and equipment required for the training. The trainer may also duplicate a copy of the standard recipe for the trainee, ensure all equipment is clean and available, and confirm that all tools, supplies, and ingredients are assembled and available.

✓ *Set-up workstation*—The trainer should ensure that the workstation is relatively free of anything that might detract from the training.

✓ *Prepare the trainee*—A new employee should know the purpose of the training: to provide the knowledge and skills needed to perform one of the tasks required for their position. Experienced staff receiving on-job training should understand that the purpose of the training is to provide the knowledge/skills needed to perform a task in a different way or to learn how to perform a new task. Trainees should be told how the training they will receive in the training lesson relates to their overall training experience addressed in the training plan.

Figure 9.1 Four-Step On-Job Training

✓ *Determine what the trainee already knows*—This is easy when on-job training is used. Assume a specific item of equipment must be operated. If the trainee claims to have this skill, they can demonstrate it. If successfully done, that aspect of training is not necessary. If the trainee cannot correctly operate the equipment, the training should focus on this task at the appropriate point. Then, the trainer can maximize available training time on steps where training is most needed.

Step 2: Presentation

Presentation is the second step in the on-job training process. To review appropriate training presentation procedures, assume a trainer is tasked with teaching a new employee how to conduct a **physical inventory** in a foodservice operation's storeroom.

The training session begins when the trainer provides an overview of the training lesson's objective and then explains the importance of the task. The trainer might, for example, indicate that "a physical inventory count helps confirm the accuracy of **perpetual inventory** records, and we do it monthly."

The training occurs in the storeroom with an overview of the task: "Two people are required. One person physically counts the number of cases, cartons, or bags of each product in inventory. A second person verifies the inventory quantity and then enters it onto a worksheet. This process is repeated until all products are counted; it usually takes about one hour to complete the physical inventory in this storage area."

Key Term

Physical inventory: A tool in an inventory control system in which an actual (physical) count and valuation of all product inventory on hand is taken at the close of an accounting period.

Key Term

Perpetual inventory: An inventory control system in which additions to and deletions from the total inventory are recorded as they occur.

After this explanation, applicable activities are demonstrated. For example, the trainer shows the trainee how the storage area is organized, and then reviews how the inventory worksheet is completed which includes the date the physical inventory is taken.

The training lesson used is well developed, so the task is divided into separate and teachable steps. The trainer explains the first step in the task, answers questions posed by the trainee, and then allows him or her to repeat, practice, and/or demonstrate the step. For example, some products are counted by the purchase unit (for example, a case or carton), while others are counted by an individual unit (for example, a can or bottle).

Each of these key points in the lesson is discussed in a separate training step. If necessary, the sequence of steps can be repeated so the trainee can learn all steps.

As the presentation process evolves, trainers follow several principles.

✓ They speak in simple terms and do not use excessive industry jargon.
✓ When possible, easier tasks are presented before more complex activities are discussed. For example, the trainer teaches how to conduct a physical count before analysis of data from computer printouts is considered.
✓ Tasks are explained and demonstrated slowly and clearly.

✓ A questioning process is used to help ensure trainee comprehension. The trainer does not use a closed-ended question ("Do you understand what I am doing?"). Instead, an open-ended question is used: "Why do you think full cases are counted before opened cases?"

✓ Clear and well-thought-out instructions for each task are provided. The trainer indicates why each step is necessary and should be done in a specific sequence.

✓ The trainer asks questions to help ensure that the trainee understands and to suggest when additional information, practice, and/or demonstration can be helpful.

Step 3: Practice and Demonstration

The third step in on-job training allows the trainee to practice and demonstrate what has been taught. During this step, several principles become important.

✓ The trainee should be asked to repeat or explain each key point.

✓ The trainee should be allowed to demonstrate and/or practice the task. If practical, the trainee should practice each step to learn its "basics" before training continues. Then, the trainer can confirm the trainee knows how to perform the task and that the trainee understands what is required for successful task performance. Steps are typically taught by the trainer and have been practiced and/or demonstrated by the trainee in the proper sequence.

✓ The trainer can coach the trainee to reinforce positive performance ("Andrew, you did that task perfectly!") and to correct improper performance ("Andrew, the next item to be counted in the inventory is listed on the inventory sheet according to its storeroom location; we can't skip ahead.")

✓ Trainers should recognize that when the task and/or steps are difficult, initial progress may be slow. Then the trainee may require more repetition to build speed and consistently and correctly perform the task or step.

✓ Trainers must realize that some trainees learn faster than others. This principle is especially easy to incorporate into training when the on-job training approach is used. Within reason, training can be presented at the pace judged "best" for the individual trainee. Also, the time allowed for trainee demonstration can be varied as needed.

✓ Correct performance should be acknowledged before addressing performance problems. Some trainers refer to this as the "sandwich method" of appraisal. Just as a sandwich has two slices of bread with a filling between them, the sandwich method uses an introductory phrase ("Phyllis, you have mastered almost all parts of this task"), followed by problem identification ("Just ensure the meat slicer is set to the correct thickness before slicing the roast beef"). This is followed by a concluding phrase. ("I'm glad you are learning how to properly cut the beef for our French Dip sandwich, and you are doing it really well!")

✓ Trainees should be praised for proper performance. It is probably not possible to say too many good things to a trainee! Everyone likes to be thanked for a job well done, to be told how important and special they are, and to receive immediate input about performance. Trainers should reward trainees for success by noting it and by thanking the trainees for correctly completing the task.

Step 4: Coaching

The coaching step includes concluding activities to help ensure the training is effective: performance-based training objectives are attained. Useful coaching procedures include the following:

✓ At the end of the training, the trainee should be asked to perform, in sequence, each step in the task.
✓ The trainer should encourage questions and ask them.
✓ The trainer should provide ongoing reinforcement about trainees' positive attitudes and when the trainees improve their skills and knowledge.
✓ Close supervision immediately after training and occasional supervision after a task is mastered can help ensure the trainee consistently performs the task correctly.
✓ The trainer may ask the trainee about suggestions for other ways to perform tasks after the staff member has gained experience on the job.
✓ Trainees should be asked to retain copies of training materials provided during the session for later referral, if appropriate.

Trainer/coaches will be most effective when they:

1) Encourage the trainee to seek assistance when needed.
2) Tell the trainee who to contact if assistance is needed.
3) Check the trainee's performance frequently but unobtrusively.
4) Reinforce proper performance by letting trainees know how they are doing.
5) Help the trainee to correct any mistakes.
6) Ensure that any mistakes that are made are not repeated.
7) Ask the trainee about suggestions for better ways to do the task.
8) Complement the trainee for successful demonstration.

Find Out More

Professional athletes know the value of coaching to the success of their teams, and excellence in coaching most often leads to improvement in a sport team's performance.

Coaching is just as important in the workplace as it is in professional sports. Those foodservice operators who excel in coaching:

1) Give their team members regular and frequent feedback
2) Create a culture of team unity
3) Push team members to attainable limits
4) Are open to team members' ideas
5) Encourage team members to learn from each other
6) Build individual team member confidence
7) Are tolerant of team member's mistakes
8) Regularly ask team members what they can do to help

A great coach will make team members see what they can be, rather than what they are. The best foodservice operators let their team members know they should come to them with questions or concerns. They use one-on-one conversations to understand the challenges the team faces and to build plans to overcome them. These operators let employees know they are there to help employees who should feel comfortable asking for advice or help.

To learn more about the ways coaches in foodservice operations can help optimize their work team's performance, enter "tips for coaching hourly employees" in your favorite search engine and review the results.

Other Individual Training Methods

The four-step on-job training method just addressed is frequently used in foodservice operations. There are, however, other individualized training processes that can be used alone or with another training method, including:

✓ *Self-study*—Trainees can enroll in various types of **distance education** programs offered by an educational institution, professional association, or other training entity. Self-study in a broader sense also occurs when interested staff members, perhaps with the advice of a supervisor or mentor, enroll in a college course, read a recommended book, or view training materials online while on the job.

Key Term

Distance education: An individual training method in which a staff member enrolls in a for-credit or not-for-credit program offered by an educational facility or a professional association.

Technology is rapidly changing how many people participate in self-study courses. Today's generation of new employees are very familiar with technology and use it for many purposes, including **e-learning**.

Benefits of e-learning include that trainees can study when and where they desire. Also, tests including evaluations before and after the course can help to assess learning and results can be electronically tabulated and made available to the trainee's supervisor. Content can be changed quickly and updates can be put in place almost immediately.

Key Term

E-learning: Learning conducted via electronic media and typically on the internet. Also commonly referred to as "online learning."

E-learning programs are typically self-paced, and trainees can learn what is relevant and avoid unnecessary and/or already known information. With many alternatives, feedback is possible because employees can e-mail questions to and receive responses from the training facilitator.

E-learning can also be used to pretest employees for placement in on-site training programs and even provide background content/information to expedite traditional training.

Technology at Work

While there are many off-the-shelf programs designed for restaurant employees working in different positions, foodservice operators often want to provide training that is specific to their own businesses.

In the past, creating customized e-learning training programs was beyond the capabilities of most foodservice operators. Today, however, advancements in technology have made it easy for many operators to create their own customized training materials.

Available foodservice training software makes it possible to develop interesting courses in menu item production and service that are unique to a single operation. It is possible to add reading materials, instructional videos, quizzes, and summary tests that engage staff members and can be easily updated to keep pace with evolving industry trends. These courses, once developed, can be accessed by staff anytime, anywhere, and on any smart device.

To review some of the software available to foodservice operators who want to create their own online training materials enter "customized training software for restaurants" in your favorite search engine and view the results.

✓ *Cross-training*—The cross-training (see Chapter 4) method of individual training consists of activities that allow staff to learn how to perform tasks in another position. For example, a dishwasher may learn how to pre-prep vegetables. This may be done at a staff member's request or to help some employees gain knowledge and skills that will be helpful when they have the opportunity to assume another position.

Cross-training can be an effective training method, and it has the added advantage of increasing a foodservice operator's scheduling and staffing flexibility if staff absences or staff resignations occur. The reason: properly cross-trained employees will be ready to assume positions that become vacant unexpectedly.

✓ *Job enlargement*—**Job enlargement** occurs when a trainee learns tasks that are traditionally performed at a higher organizational level. As with other individual training methods, this tactic can benefit both the trainee and the organization. It can be especially useful when a supervisor or lead team member must take on additional responsibilities, and, to do so, must delegate some current duties.

✓ *Job enrichment*—**Job enrichment** occurs when additional tasks that are part of a position at one organizational level are added to another position at the same level. For example, assume a cook's position is at the same organizational level as that of a baker. Job enrichment could occur as the cook learns baking tasks and as the baker learns cooking tasks.

✓ *Job rotation*—**Job rotation** is an individual training method that involves the temporary assignment of employees to different tasks to provide work variety or experience. As with other individualized work methods, job rotation benefits the staff member and helps to create back-up expertise within the foodservice operation.

Key Term

Job enlargement: An individual training method that involves adding tasks to a position that are traditionally performed at a higher organizational level.

Key Term

Job enrichment: An individual training method that occurs when additional tasks that are part of a position at the same organizational level are added to another position at the same organizational level.

Key Term

Job rotation: The temporary assignment of employees to different positions or tasks to provide work variety or experience and to create "back-up" expertise within the organization.

Each individual training method noted above can help a foodservice operation over the short and longer term. However, each can also be part of a formal effort to provide additional training to staff members who are judged to be proficient in their existing position. There are numerous methods of involving staff in ongoing training, and these training activities are applicable to foodservice operations of all sizes.

What Would You Do? 9.1

"I told her twice how to do it, so it's not my fault," said Lamont, the head bartender at the Bottle and Barrel Brewing Company.

Lamont was talking to Shera Teagues, the owner of the Bottle and Barrel. Shera's brewery/pub was one of the most popular craft breweries in the city. Between the interior décor, the music, the food, the staff, and importantly, the craft beers she offered, Shera had built a thriving business catering to a younger clientele of craft beer aficionados.

Shera and Lamont were talking about a problem that occurred the previous night. Catherine, a new bartender, had been assigned to closing the pub. As a result, she was working alone for about half an hour before closing when one of the kegs of a popular locally crafted beer became empty. There was an additional keg of the same product available, but the problem, as they were to find out, was that Catherine did not know how to replace the keg. That upset several guests who did not get the product they wanted.

"What did you say to her when you told her how to do it?" asked Shera.

"It was the first day she got here. I said to her, it's simple. First, you turn the gas off then lift the coupler handle up. Once the handle is up, you twist the coupler a quarter turn counter-clockwise, and then pull it straight up. Now simply locate the keg you want to put on the tap and remove the plastic cover. Slide the coupler into place, turn it clockwise to lock it in place, then depress the handle and turn the gas back on," replied Lamont, "Like I just said, I told her how to do it, and it's not my fault she didn't listen!"

Assume you were Shera. Do you think that on her first day, telling this new employee how to change a keg, rather than demonstrating and allowing the employee to practice doing it, was an effective training method? What should you do now to prevent problems like this from happening in the future?

Group Training

Group training is used to teach the same job-related information to more than one trainee at the same time, and it can be done at or away from the work site.

Advantages of group training are that it can be time- and cost-effective when more than one

Group training: A training method that involves presenting the same information to more than one trainee at the same time.

staff member must receive the same information (for example, several cooks need to learn how to prepare a new menu item). Group training can also allow a large amount of information to be provided to many team members in a relatively short amount of time.

Potential disadvantages of group training typically relate to the difficulty of or inability to consider the individual differences of trainees when they are trained. It is typically not possible to focus on a specific trainee's knowledge and experience, speed of learning, or desire to receive immediate and individualized feedback if team members are trained in a large group.

Fortunately, the advantages of both individual and group training can often be incorporated into one training program. For example, all new staff members can learn general information in a group orientation program (see Chapter 7). This topic can be followed with individualized training so each trainee can learn specific tasks required for their own position.

There are two popular group training methods most often used in foodservice operations:

✓ *Lecture method*—In this method, the trainer talks and may use audio/video equipment or handouts to facilitate the session, and question/answer components may also be included. This method provides much information quickly, but it does not typically allow active trainee participation.

✓ *Demonstration method*—In this method, the trainer physically shows trainees how to perform position-specific tasks. Trainees can hear and see how something is done, often in the actual work environment. A potential disadvantage is that, without immediate and frequent repetition, trainees may forget some of the basic information presented. The best group training methods often include some lecture supplemented with other activities such as demonstrations, question and answer sessions, and practice time.

Preparing for Group Training

Experienced trainers know that the best training content does not "automatically" yield the best training program. An example: problems can occur if the planning process does not address the training environment, including the training tools needed to facilitate the group training.

Group training rooms should be clean, well-vented, free from noisy distractions, and provide a controlled room temperature. Unfortunately, in many foodservice operations, these types of spaces are not available. Instead, trainers must use multi-purpose space such as the operation's dining room, staff dining areas, or, in some cases, even production or other work areas.

Proper table/chair arrangements help facilitate training. Areas at the front of the room allow space for the trainer's materials and equipment. This can include a table, lectern, flip chart(s), laptop computer, and digital projector. Other items include those necessary for demonstrations, handouts, or other needs. Trainers also appreciate ice water or another beverage, so table-top space for this purpose is helpful if it can be provided. Other considerations when planning training room set-ups require that there be good line-of-sight opportunities so all trainees can see the trainer and any display items (video monitors, screens, flip charts, chalkboards, and the like).

Handouts can sometimes supplement and enhance group training. They may contain a brief training outline or an exercise to be completed after there is applicable discussion. Trainers should consider when handouts should be circulated (for example, before or at the beginning of the training session, or when they are discussed). They should also consider when and how much time is required for trainees to read the handouts before they are used.

Delivering Group Training

Effective trainers must be good communicators. This is important when an individual on-job training method is used, but it is especially critical for group training. When training groups, trainers have fewer opportunities to solicit feedback and determine whether training content is understood. The effective delivery of group training begins with a well-planned training lesson or outline that identifies the main points and subpoints to be addressed, the time scheduled for the entire session, and for the training points within it.

Basic procedures for delivering effective group training content include:

✓ *Using a warm and genuine introduction.* If necessary, the trainer should introduce himself/herself and allow the trainees to introduce themselves. Personal introductions can include information about each trainee's position, years of experience, and personal and professional training goals.

✓ *The introduction should focus on "what's in it" from the trainees' perspectives.* For example, "lately we experienced some problems in taking inventory that have caused frustrating rework. Thanks to your input, we've collected some ideas to resolve these problems, and we're going to discuss them today."

✓ *The training session can be previewed.* Training goals can be noted and trainees can be informed about training tactics. For example, "today we will talk about the inventory process and then conduct a role play exercise that will be fun and interesting."

✓ *Indicate when questions should be asked:* Trainers should indicate if questions can be asked when they arise during the session or if they should be held for discussion at its end.

✓ *Gain the trainees' attention.* Trainers can ask for personal stories about situations related to the training topic.

Key aspects of group training delivery include:

✓ *Limiting the number of training points.* This is easy when an effectively developed training lesson provides the presentation's outline.
✓ *Discussing topics in the same sequence as noted in the introduction.* If, for example, three objectives were previewed at the beginning of the session, they should drive the training sequence that follows.
✓ *Using transitions between main points.* For example, a trainer could say "Now that you know how to assemble the equipment, let's learn how to safely operate it."
✓ *Concluding each main point with a summary.* For example, a trainer could say "To this point, you have learned that the equipment must be operated according to the manufacturer's instructions. Now let's learn how to use it for some specialty items we make here."
✓ *Managing questions effectively.* Trainers can repeat the question, provide a response, or ask other trainees to provide their comments. After discussion, requesting that the trainee who initially asked the question respond to an open-ended question can help the trainer ensure the trainee understands.
✓ *Being aware of nonverbal communication.* Body language can frequently tell the trainees more than the trainer says. For example, facial expressions can suggest that the trainer is frustrated with their inability to learn, and pacing back and forth may suggest the trainer is bored.

Experienced trainers know that sometimes more, rather than less, time than planned is needed to fully address a training concept. To compensate, they try to stay on point. They also evaluate training sessions as they evolve to make time adjustments that focus on the most important training points. Also, they speak at an appropriate volume and do not speak too quickly or slowly. Effective trainers avoid slang terms and jargon that can hinder communication effectiveness.

In some cases, the trainees' language fluency can become a significant concern that adds a new dimension to many of the points already made about effective group training presentations. Effective trainers pronounce words correctly, change voice tones, and avoid public speaking errors such as the frequent use of expressions such as, "you know," "let's see," "ahhhhh," and "like, you know."

Find Out More

Many foodservice operators find that using simple role play exercises (see Chapter 8) in small group (4–6 participants) training is an effective way to identify and provide solutions to typical problems encountered by staff. When properly planned and implemented, incorporating role-playing in small group training can be fun and productive.

(Continued)

For example, a role play could be set up to instruct servers on how best to handle "rude" customers. Also, a role play exercise may be an excellent way to identify the best server response(s) to a guest's question of "What do you recommend?"

Role-playing can be effective for back-of-house issues as well. For example, how should kitchen staff react when they accidentally send out a wrong order and what steps should they take to correct the issue? These can be illustrated in a role play.

An effective role play activity might only last for 5–10 minutes with the trainer setting up the scenario, and trainees taking the role of "actors" within it. Sometimes the trainer may even be one of the role players, and they can encourage comments and ideas from trainees who are among the actors.

To see other suggestions of good ideas for role-playing activities that can be used, enter "how to incorporate role playing in restaurant training" in your favorite search engine and view the results.

Special Training Considerations

Regardless of whether a foodservice operator uses individual on-job training, group training, or a combination of both, it is important to recognize that there are several training-related considerations to consider. Among the most important are:

✓ *Addressing annoying or distracting mannerisms*—A trainer's body language can distract trainees. Some trainers may not be aware of these expressions and actions and could easily correct them if they knew about them. Trainers can make a presentation to professional colleagues and solicit objective feedback, or they can arrange a videotaping to learn something about their presentation style that might otherwise be overlooked.

✓ *Scheduling training so tasks taught can be implemented on a timely basis*—When properly prepared, trainees will be enthusiastic about their training and want to immediately begin applying the knowledge and skills they have learned. This may not be possible when, for example, necessary equipment has not been installed or when one phase of an initiative must be implemented before a follow-up phase. Explanations about the need to delay training and/or to provide reasons for these challenges can be helpful.

✓ *Effectively managing trainees who do not appear to want training*—For a variety of reasons, some trainees may resist participating fully in a training session. If that occurs, trainers can make a direct request ("Steve, we're looking for input from everyone after this discussion, so you'll need to be prepared for it").

Taking a short training break so the trainer can personally speak with an inattentive trainee may also be useful. There may be other times when many trainees appear unresponsive. Then reconsidering the training approach and implementing improvement efforts to reduce these types of concerns may be necessary.

✓ *Not attempting to accomplish too much*—The best trainers ensure that they allow sufficient time for practice (skill training) and discussion (group training). They realize that effective training may take more than the originally planned training time.

✓ *Listening to trainees*—Trainers must be alert for trainee "overload" and side conversations, blank stares, and a preoccupation with non-training activities. The trainees' body language can sometimes "speak" volumes to trainers about their training efforts.

✓ *When needed, segmenting the training*—For example, assume that new technology will be used to manage inventory data. Should that technology be installed in the quickest manner possible, and should all affected staff members be trained accordingly? Alternatively, is a phased implementation of the new technology with staggered training most appropriate? Experienced trainers will carefully consider these issues and make decisions based on their analysis.

✓ *Keeping the training focused on stated objectives*—This is easily done when training plans and training lessons have been carefully developed and are consistently used. The trainer's role is to facilitate training by pacing it according to a realistic schedule.

✓ *Not "reinventing the wheel"*—If cost-effective off-the-shelf training materials are available, then they should be used. A corollary to this principle is also important: don't rewrite training materials that have been developed in previous sessions. The use of an organized training handbook (see Chapter 8) can ensure that required materials can easily be located when needed.

✓ *Rehearsing before training*—Just as trainees must practice learning a task, trainers should also practice to gain experience with their presentations. This could be done alone or in front of a small group of supervisors or managers who can properly critique the trainer's delivery.

✓ *Avoiding the development of a trainer's ego*—It is important that trainers always remember that the objective of training is to improve the trainees' performance, not to impress trainees with the trainer's knowledge, skills, or experience.

✓ *Keeping training sessions as short as possible*—It may have been a long time since some trainees participated in a training session. Many will feel more at ease working on the job than being in a training session. Experienced trainers also know that several short training sessions are generally better than one relatively long one.

✓ *Recognizing the importance of the training environment*—Experienced trainers know that the environment of the training area (meeting room or work site) is often more detrimental to training effectiveness than is the training content or its delivery.

✓ *Respecting trainees' knowledge and experience*—Wise trainers recognize that adult trainees bring a wide variety of personal experiences, attitudes, core values, and preconceptions to their training experience. They recognize and use these experiences to deliver the best possible training.

✓ *Linking training to assessment and performance*—In most cases, the goal of training is to create behavioral or attitudinal change, and this will yield a result (effect) that is beneficial to the trainees and the organization.

✓ *Having fun!*—Trainees cannot learn properly if they do not enjoy their training. What is pleasurable to one trainee may not be for another. When possible, trainers should use a variety of training tactics that involve active trainee participation because these are preferable to less interactive approaches.

Evaluating Training Efforts

Evaluation is the final step in the employee training process, and it can indicate if the training was successful (were goals attained?). Alternatively, the evaluation can suggest the need to repeat the entire planning process or to refocus efforts on specific training steps.

Foodservice operators' time, money, and labor are increasingly in limited supply. As a result, they must determine whether their commitment of resources to planning and implementing training procedures is a better use than are other alternatives. This is one reason training evaluation is important. Additional reasons to evaluate training efforts include to:

✓ *Assess the extent to which training achieved planned results*—Training objectives identify competencies to be addressed in training, and they provide a benchmark against which training can be evaluated. Assume one objective of a training program is to teach a server how to enter guest orders in an operation's POS system and to adjust the guests' check as changes to the order are made. A second objective is to show the server how to tally the check for presentation to the guest and the proper method to collect payment using 10 sequential steps. These two objectives drive the training (how to use the POS system), and they provide a way to evaluate training effectiveness (were each of the required steps properly completed?).

✓ *Identify strengths and weaknesses of training*—Few training programs are 100% effective or ineffective. Some training lessons are better than others, some training activities are more useful than their counterparts, and some trainers may be more effective than their peers. Successful evaluation can identify training aspects that should be continued and elements that may require revision.

✓ *Determine the success of individual trainees*—Trainees who are successful (they achieve planned results on the job) will not require remedial training. However, other trainees may need revised and/or repeated training. Assessment of individual trainees is relatively easy when an individualized, on-job training method is used, but it is more difficult with group training. The importance of the assessment is, however, equally important.

✓ *Gather information to help justify future programs*—When the success of a training activity is quantified, objective information becomes available to help justify future training efforts. Alternatively, foodservice operators can determine whether resources are better invested in other facility improvement efforts.

✓ *Determine trainees eligible for future training*—Some foodservice organizations provide educational or training activities with formalized **career development programs** that require prerequisite training.

Key Term

Career development program: A planning strategy in which one identifies career goals and then plans education and training activities designed to attain them.

Other foodservice operators have formal or informal "fast-track" programs in which selected trainees who successfully complete training programs are eligible for additional training opportunities. These, in turn, lead to increased promotional considerations as vacancies occur. In both of these instances, operators must know whether and to what extent individual trainees successfully completed the training.

✓ *Assess costs and benefits of training*—The expenditure of any resource must generate a return greater than the cost of resources allocated for it if the funds spent are judged to be cost-effective. Some benefits of training, including improved morale and increased interest in attaining quality goals are generally difficult to quantify. Others, including improved guest service skills and reduced operating costs for a specified task, may be easier to quantify, and both could be assessed by training evaluation.

✓ *Reinforce major points for trainees*—Some training evaluation methods, including written quizzes or tests, and performance appraisal interviews allow trainers to reinforce their most important training points. For example, if questions on a written assessment address the most important training concepts, they can be self-graded or reviewed by the trainer, and reinforcement of these important points is possible.

✓ *Assess trainees' reactions to training*—Trainers who are interested in improving training programs usually want to gain their trainees' perspectives about the programs. When possible, anonymous input gained before, during, and after training can be helpful in this assessment.

✓ *Assess trainers' reactions to training*—"There's always a better way" is an old saying that applies to training as well as to other management tactics. Trainers who have used training lessons, for example, may well have ideas about ways to improve them in the future.

Some trainers think about training evaluation only in the context of an after-training assessment. While training should be evaluated at its completion (and perhaps again several months after its completion), evaluation can also be helpful before training even begins and while it is conducted.

Effective trainers understand that training assessment tools can work best when the methods to evaluate training are:

1) **Valid**—Training evaluation methods must be valid: they must measure what they are supposed to measure. For example, assume a training objective focuses on the ability of trainees to prepare a new Caesar salad entrée that was taught during the training. The trainees' ability to follow the proper standardized recipe procedures suggests that the training was successful (at least at the time of the demonstration). In contrast, if training assessment queried trainees about issues such as "Did you like the training?" and "Did the trainer seem enthusiastic?" the trainers could not determine whether training objectives were attained.

2) **Reliable**—Training evaluation methods are reliable when they consistently provide the same results. Training activities that are implemented in the same way by the same trainers using the same training resources and procedures for employees in the same position may be consistent. Will the results be the same or similar each time it is replicated? Trainers do not know unless the same evaluation methods are used.

3) **Objective**—Objective evaluation methods provide quantitative (measurable) training assessments. Acquisition of knowledge can be objectively measured by performance on a well-designed test. Efficiency in a skill might best be assessed by observing the trainee's performance of the task after training. Then performance can be considered "acceptable" if the procedures used are the same as those taught during training.

4) **Practical**—A training evaluation method is practical when the time and effort required for the assessment are "worth" its results. Knowledge assessments that require trainees to memorize mundane facts and skill demonstrations that are benchmarked against staff with extensive experience and efficiency in performing the task are not helpful.

5) *Simple*—An evaluation method is simple when it is easily utilized by the trainer, when the trainees easily understand it, and when results are easy to assess and analyze by those evaluating the training.

When evaluation methods have the above characteristics, they may be selected for use in training program evaluation before, during, and/or after training is delivered.

Evaluation Before Training

To understand the importance of evaluation before training, assume that a trainee participated in a food safety training session and, after the training session was completed, a written test was administered. Also, assume that the trainee missed only two questions of the 20 that were asked. Many trainers would likely conclude that the training was successful because the participant scored 90% (18 questions correct ÷ 20 questions total = 90%).

In fact, the training may have wasted the operation's resources and the trainee's time if the trainee *already* knew the information taught before the session began. In this example, the after-training evaluation really measured what the trainee knew when the training was completed rather than what was learned directly from the training session.

To address this concern, some trainers use a **pretest/post-test evaluation** tool in which key concepts to be addressed during the training are identified, and these concepts are addressed in a pretest administered before the training begins.

Then the same measurement tool using the same questions is administered at the end of training. The improvement (change) in scores between the pretest and post-tests represents one measure of training effectiveness.

There are other advantages of pretest/post-test evaluation:

Key Term

Pretest/post-test evaluation: A before and after assessment used to measure whether the expected changes took place in a trainee's knowledge, skill level, and/or attitude after the completion of training.

✓ It provides trainees with an overview (preview) of the training.
✓ It helps trainees identify some of the most important concepts to be addressed during training.
✓ It presents an opportunity for trainers to preview the lesson and suggest priority learning points before training begins.

As is true with other training evaluation methods, administering the post-training assessment several months (or even longer) after training is completed may also be helpful. This tactic can help determine the extent to which training information has been retained for use in the workplace.

Evaluation During Training

Some types of training evaluation can occur during the training session itself. For example, trainers can use an introductory session statement to state that they will ask for feedback during the session. When this feedback is solicited, trainers can obtain a reality check and perhaps learn helpful information to improve the remainder of the training.

Trainers facilitating group sessions can ask trainees to write anonymous responses to statements such as:

✓ I wish you would stop doing (saying) . . .
✓ I hope we continue to . . .
✓ I don't understand . . .
✓ I hope you will begin to . . .
✓ A concept that I wish you would discuss further is . . .
✓ A concept I want to learn more about that has yet to be discussed is . . .

The major goal of "during the training" evaluation is to learn how to maximize use of the remaining training time. Then revisions to training content and/or delivery methods can better ensure attainment of training objectives.

Evaluation After Training

After-training evaluation can assess the extent to which training achieved its planned results, and it may also identify how training sessions might be improved. Several after-training evaluation methods are in common use. One or more of these methods can provide information for foodservice operators to improve future training efforts.

Common after-training assessment methods include:

✓ *Observation of after-training performance*—Owners, operators, and supervisors can "manage by walking around," and, in the process, note whether knowledge and skills taught during training are being applied. For example, storeroom personnel can be observed as they receive incoming products, and procedures used can be compared with those presented during training. Note that when proper procedures are used, a "Great job!" compliment is always in order. By contrast, a coaching activity to remind staff members about incorrectly performed procedures may also be needed.

✓ *Objective tests*—These can be written, oral, and/or skill-based and include traditional written exams and quizzes or after-training demonstrations. Written examinations including multiple choice and true/false questions are most often used in foodservice operations because they are **objective tests**.

In an objective test, there is only one correct answer, little or no interpretation is needed, and minimal time is required for trainees to complete the exam and for trainers to score it. Analysis of incorrect answers can often lead trainers to identify areas in which training emphasis were insufficient, and/or trainee understanding was deficient.

✓ *Third-party opinions*—Feedback from guests can help assess training programs that addressed aspects of products and services that affected them. The use of a **mystery shopper** in some types of foodservice operations is another example. Feedback can also be generated by comment cards, interviews, and/or follow-up surveys with guests in person, or online.

Key Term

Objective test: An assessment tool such as a multiple choice or true/false test whose questions have only one correct answer and yield a reduced need for trainers to interpret trainees' responses.

Key Term

Mystery shopper: A person posing as a foodservice guest who observes and experiences an operation's products and services during a visit and who then reports findings to the operation's owner. Also referred to as a "secret shopper."

Technology at Work

Mystery (secret) shopper programs are very popular and common in the foodservice industry. Foodservice operation owners often select the services of mystery shoppers because they know the reports delivered by them will be unbiased.

Those entities supplying mystery shopping services typically offer a wide range of reporting options. Foodservice operators selecting mystery shopper services can choose those reports they feel will be of most value to them.

Traditional mystery shopper programs relied primarily upon written reports reflecting the mystery shopper's experience. Today, however, increasing numbers of mystery shoppers are using video to demonstrate the guest experience. For example, rather than simply write a report on restroom cleanliness, the mystery shopper can videotape what was seen and experienced when they visited an operation's restroom.

To see examples of the type of services and the written and video-based reports offered by mystery shoppers, enter "mystery shopper services for restaurants" in your favorite search engine and review the results.

✓ *Analysis of operating data*—In some cases, an assessment of operating data can yield valuable information about the effectiveness of training efforts. For example, training that addresses guest service and food costs should yield, respectively, increased guest service scores and lowered food costs (if components of these data can be separated) to determine how they were influenced by training. Training designed to help servers sell specific entrees or desserts, for example, could be measured by examining pre- and post-training sales data for these items.

✓ *Analysis of user-generated content (UGC) site scores*—UGC scores (see Chapter 8) are an increasingly critical factor in the successful marketing and operation of a profitable foodservice business. Just as UGC site scores can be reviewed to indicate potential areas where training is needed, training lessons developed to address the issues identified on UGC sites should result in increased scores.

Improvements in UGC site scores can be readily measured, and today these scores are perhaps the most significant method of determining the success of an operator's training efforts.

Regardless of the metric(s) used to assess an operation's training efforts, post-training evaluation is a key part of an operation's overall training effort, and it should not be neglected.

It is important to remember that documentation is an essential part of the training effort. Training records to be maintained in a trainee's personal file include:

✓ Name of trainee
✓ Training dates
✓ Training topics
✓ Notes, if any, regarding successful completion
✓ Other applicable information

This documentation is useful for planning professional development programs, for considering staff member promotions, and for performance appraisal (see Chapter 11). Documentation of training is also helpful when trainers develop long- or short-term plans that address training and professional development opportunities for staff members.

In most cases, when an operation's staff are well-trained and comfortable in their jobs, pay will not be the central issue responsible for attracting and retaining excellent employees. In general, however, if staff members feel they are unfairly paid, or if they are scheduled to work in ways that are perceived to be unfair, they may seek other jobs they believe will more equitably reward their efforts. The importance of pay and employee scheduling to the long-term retention of capable staff are closely related, and these two key issues will be the topic of the next chapter.

What Would You Do? 9.2

"I know it sounds kind of complicated to make, and it is, but I think our guests are gonna love it!" said Myron Coban, the manager of Love You A Latte, a coffee shop in town.

Myron was talking to Janice Fielder, the shop's owner. The holidays were coming up, and Myron wanted to feature a new item he had named the "Light Peppermint Mocha Frappuccino" as a holiday special.

"O.K.," said Janice as she sipped the new iced coffee drink Myron had prepared for her. "I agree that it's good. What did you say is in it?"

"Coffee, half and half, cocoa powder, chocolate syrup, peppermint extract, crushed peppermint candies, brown sugar, and whipped cream," replied Myron, "the secret is in knowing how much of each ingredient to use and how to blend and layer them properly."

"Well, counting all of our full and part timers we have 30 baristas working for us," said Janice, "this drink is great, but it sounds pretty complicated to make. How are you going to make sure all our staff know how to make this in a way that will provide a consistently excellent product?"

Assume you were Myron. Do you think your baristas would best be trained to make this new menu item using individual on-job training, group training, or a combination of both? Who should do the training? Explain your answers.

Key Terms

On-job training (method)
Coaching
Standardized recipe
Physical inventory
Perpetual inventory

Distance education
E-learning
Job enlargement
Job enrichment
Job rotation
Group training

Career development program
Pretest/post-test evaluation
Objective test
Mystery shopper

Operator's 10-Point Tactics for Success Checklist

Evaluate your need for, and the current status of, each of the following operational tactics. For those tactics you think are important, but not yet in place, develop an action plan for its implementation including who will be responsible for the tactic's completion and the target date by which it should be completed.

Tactic	Don't Agree (Not Done)	Agree (Done)	Agree (Not Done)	Who Is Responsible?	Target Completion Date
				If Not Done	
1) Operator recognizes the value of on-job training to the development of a productive work team.	___	___	___		
2) Operator can summarize key components of "preparation," the first step in on-job training program development.	___	___	___		
3) Operator can summarize key components of "presentation," the second step in on-job training program development.	___	___	___		
4) Operator can summarize key components of "practice and demonstration," the third step in on-job training program development.	___	___	___		
5) Operator can summarize key components of "coaching," the final step in on-job training program development.	___	___	___		
6) Operator can state the methods used to properly prepare for effective group training.	___	___	___		

				If Not Done	
Tactic	Don't Agree (Not Done)	Agree (Done)	Agree (Not Done)	Who Is Responsible?	Target Completion Date
7) Operator can state the methods used to properly deliver effective group training.	——	——	——		
8) Operator recognizes the importance of evaluating training efforts before training.	——	——	——		
9) Operator recognizes the importance of evaluating training efforts during training.	——	——	——		
10) Operator recognizes the importance of evaluating training efforts after training has been completed.	——	——	——		

10

Compensating and Scheduling Staff

Operator's Brief

In this chapter, you will learn about the importance of carefully constructing your foodservice operation's total compensation package and how to create a master employee work schedule.

The compensation offered to employees can be viewed from several perspectives including that of a business owner, its operators, and individual staff members. In this chapter, you will learn how to fully understand each of these perspectives when developing a comprehensive compensation package that provides extrinsic and intrinsic rewards to your employees.

The development of a compensation package must consider the laws and regulations to be followed when paying employees. Some of these regulations are established by the federal government while others may be established by a state or a local regulatory agency.

In most foodservice operations some employees are salaried, and others will be paid an hourly wage. In this chapter, you will learn how these two approaches to employee pay are managed. In addition, since many foodservice operators employ workers who regularly receive tips, special rules related to compensating tipped employees are also addressed.

While the amount of money paid to your employees for their work is important, the voluntary and involuntary benefit packages you offer are also critical. In many cases, the total benefits package offered to staff by your operation will significantly impact your ability to attract and retain highly qualified workers.

Foodservice workers are extremely interested in their paychecks, but they are also concerned about the work schedules to which they are assigned. In the case of hourly workers, the number of hours and the specific days and times they are scheduled to work are almost as important as the hourly pay they receive. In this chapter, you will learn the steps required to successfully create, distribute and, if needed, modify your employee schedules.

CHAPTER OUTLINE

The Importance of Compensation
    The Business Owner's View of Compensation
    The Foodservice Operator's View of Compensation
    The Staff Member's View of Compensation
Compensation of Team Members
    Legal Aspects of Compensation
    Payment of Salaries
    Payment of Hourly Wages
    Compensation of Tipped Employees
    Benefit Payments
Scheduling Team Members
    Legal Aspects of Scheduling
    Schedule Development and Distribution

The Importance of Compensation

Key Term

The proper **compensation** of foodservice staff is critically important to both employees and employers because it affects so many other business issues.

Compensation: Any form of cash and non-cash payment given to an individual for services rendered as an employee by their employer.

If the staff of a foodservice operation feel they are being unfairly or inequitably compensated, they will likely seek jobs with other employers who they believe will treat them more fairly. The result can be excessive employee turnover rates and increased costs for recruiting, hiring, and training staff.

Alternatively, employers who pay their employees significantly more than other employers may find their operating costs are too high to allow them to remain competitive and achieve the profits required to stay open for business.

Unfortunately for foodservice operators, elusive concepts such as "fair," "equitable," and "competitive" pay rates defy unanimous agreement. As a result, a challenge confronting many operators is to design and manage compensation programs that are, at the same time, considered reasonable by employees and good for their own businesses. Employers and employees alike could agree that compensation should be reasonable, and the term "reasonable" can be considered the value that would ordinarily be paid for the same type of services by similar businesses under similar circumstances. The best foodservice operators, however, go one step further and use their compensation programs as important tools to attract and retain excellent workers and to maximize operating profits.

Foodservice operators designing effective compensation programs must understand the legal aspects of doing so. However, more is at stake than just ensuring a compensation program's legality. Managing compensation programs requires addressing two main issues: controlling costs and using pay to attract and retain the best staff members. An ideal compensation and benefits program tracks costs, ensures pay equity, is understood by all employees, and recognizes the long-term wisdom of balancing the financial interests of employees and employers.

To best understand the development of a proper and effective compensation program, one must first recognize the various views of worker compensation. These are:

✓ The Business Owner's View of Compensation
✓ The Foodservice Operator's View of Compensation
✓ The Staff Member's View of Compensation

The Business Owner's View of Compensation

From the perspective of a foodservice business's owner, the cost of labor is often exceeded only by the cost of the food and beverage products that are sold. In some cases, labor costs actually exceed food and beverage product costs. As a result, most foodservice operation owners are keenly aware that their labor costs directly affect the profit they will generate. However, some short-sighted operators believe it is best to minimize labor costs and the total amount of compensation provided to their staff.

More enlightened owners, however, recognize that their goal should never be to minimize compensation, but rather to optimize it. Experienced owners of foodservice businesses know it is not in their long-term best interests to serve their guests the cheapest possible food and beverage products. In much the same way, they also know that paying the lowest possible compensation will not usually result in the acquisition and retention of a highly skilled and motivated team of workers.

Foodservice business owners who have been successful over the long term recognize that a well-compensated staff is an important key to satisfying guests, maximizing their return business, and generating desired profit levels.

Find Out More

Foodservice owners are often frustrated by the labor shortage. To address this challenge, some owners have tried to increase productivity among existing staff, while 33% say they've reduced their hours of operation. Some owners claim that today's workers are lazy or simply "don't want to work."

Unlike what these owners believe, however, competition with unemployment benefits, and a general unwillingness to work are not the primary reasons for the restaurant worker shortage. Survey data* shows that restaurant workers are leaving the industry (or declining to join it!) because they want three things:

1) *Higher wages:* A survey of hourly workers found that 55% of employees are looking for an opportunity to increase their pay, and that higher wages (and benefits) are a top demand among restaurant workers and play the biggest role in long-term retention.

2) *Greater manager recognition:* Surveys have found that foodservice employees desire long-term career prospects and management recognition such as promotions. Opportunities for professional development and career growth are critical to employee satisfaction and, subsequently, to lower turnover rates.

3) *More schedule flexibility:* Restaurant workers often must deal with unpredictable and inflexible schedules, and this creates difficulty in striking a positive work-life balance. Only 29% of restaurants are using scheduling software to manage labor costs and shifts. However, 56% of employees say that flexible scheduling would greatly affect their happiness at work and their desire to stay in their roles.

While the challenges related to worker shortages are real, they can be overcome. To learn more about how owners of foodservice operations can address worker shortages, enter "How to attract hourly paid restaurant workers" in your favorite search engine and review the results.

*https://www.careerplug.com/blog/restaurant-worker-shortage/ (Retrieved June 7, 2023).

The Foodservice Operator's View of Compensation

For foodservice operators who are not owners, worker compensation levels are important because they are one measure of the operator's effectiveness. The most used measure of an operator's ability to effectively manage labor costs in the food service industry is the **labor cost percentage**.

Key Term

Labor cost percentage: A ratio of overall labor costs incurred relative to total revenue generated.

The formula used to calculate a labor cost percentage is:

$$\frac{\text{Total cost of labor}}{\text{Total revenue (sales) generated}} = \text{Labor cost percentage}$$

A labor cost percentage allows an operator to measure the relative cost of labor used to generate a known quantity of sales. In most cases, lower labor cost percentages are more desirable than higher labor cost percentages. The control and management of an operation's labor cost percentage will be addressed in detail in Chapter 12. For now, it is important to remember that, as an operation's total cost of labor (total amount of worker compensation paid) increases, the operation's labor cost percentage will increase as well (assuming that a fixed level of revenue is generated).

While several factors influence what may be considered a "good" labor cost percentage, in nearly all cases the amount an operator spends on compensation serves as an indicator of that operator's business skill and ability.

Salary increases, bonuses, promotions, and even the long-term employment of an operator are often dictated in part by the operator's ability to maintain a proper labor cost percentage. Therefore, foodservice operators who are not owners are keenly interested in the amount of compensation paid in their operating units.

The Staff Member's View of Compensation

While owners and operators view compensation in terms of its impact on profitability and performance evaluation, the compensation views of an operation's staff members are more complex. This is true whether the staff member is a **salaried** or hourly paid employee.

Salaried employees are, of course, keenly interested in the amount they are paid. However, like hourly paid workers, they may also assess the fairness of their salaries based on what they believe others doing the same work are paid.

To illustrate, consider a salaried staff member who is satisfied with their pay. However, in conversation with peers working in a similar operation and doing similar work, the salaried worker discovers their salary is significantly less than that of peers. In this case, it is highly likely that this salaried worker will go from being satisfied with the pay level to being unsatisfied with it, even though the amount of compensation has not changed.

Key Term

Salaried employee: An employee who regularly receives a predetermined amount of compensation each pay period on a weekly or less-frequent basis. The predetermined amount paid is not reduced because of variations in the quality or quantity (amount of time) that a salaried employee works.

In a similar manner, consider that an hourly worker making a certain rate of pay per hour may be satisfied with that amount. However, should that worker find that another person in the same operation is doing identical work and making a significantly higher hourly rate, dissatisfaction is likely to occur. The same can be true with pay increases. A worker earning an hourly rate who has been given a $0.50 per hour raise may be quite happy with the increase. But if that same worker finds that another worker doing similar work received a $1.00 per hour increase, that initial happiness may quickly go away.

In some cases, of course, there may be reasons for differences in pay even among workers doing the same jobs. For example, a worker with higher levels of seniority might make more per hour doing the same job than a worker with lesser amounts of seniority.

It is also true that pay rates in some foodservice jobs are relatively low. The result may be employees who are hard workers but still earn incomes that can make it difficult to afford food, housing, and other essentials. In these cases, employees may internalize negative stereotypes that imply they are confronted by their financial difficulties because they are incompetent or deeply flawed as people. This low self-esteem can lead to a lack of confidence and even a sense of hopelessness.

The main point for foodservice operators to remember is that compensation for staff members, whether paid a salary or an hourly rate, goes far beyond economics only. Feelings of self-worth, job satisfaction, and esteem among their peers are all factors that staff members might consider when they assess their compensation packages. It is for this reason that foodservice operators must design total compensation packages that are not only "fair," but that are also *perceived* as "fair" by their staff members.

Some inexperienced foodservice operators mistakenly feel that it is in their best interest to keep some aspects of their compensation programs a "secret" from their employees. Their rationale is that foodservice operators may want to pay some workers differently than others, and it is best if that assumption is not known by all staff members.

Experienced foodservice operators, however, recognize that the Equal Pay Act (EPA; see Chapter 2) prohibits employers from paying women and men different wages when the work performed requires equal skills, efforts, and responsibilities. Further, the EPA makes it illegal for employers to try to prevent employees from sharing pay information with each other.

The EPA makes it unlawful for an employer to have a work rule, policy, or hiring agreement that prohibits employees from discussing their wages with each other or that requires workers to obtain the employer's permission to have such discussions. As a result, foodservice operators should assume pay rates *will* be shared and compared by staff members, and this is another reason why **compensation packages** should be designed with fairness to all.

Key Term

Compensation package: The sum total of the money and other valuable items given to an employee in exchange for work performed.

The majority of foodservice workers like their jobs and enjoy the rewards they receive from working in the industry. For most of these workers, however, a critically important part of their job satisfaction relates to the compensation package they receive.

While some foodservice employees consider their jobs to be fun, few people have the luxury of working just for this reason. In most cases, workers evaluate the compensation offered when they assess the value of what they are paid for their work, and when they consider whether that pay is adequate and fair. It is important that foodservice operators ensure all employees know about their pay rates. However, it is just as important that employees be informed about other aspects of their compensation package, including items such as free or reduced cost meals, benefits, tips, and bonuses.

Most foodservice staff members would like their compensation package to be as large as possible, and many employers would like employee compensation packages to be as small as possible to maximize profits. However, experienced operators know that it is rarely in their best interest to minimize the size of compensation packages.

The reasons are twofold. First, employers who advertise positions offering below-average compensation packages tend to attract workers with lesser skills because those with greater skills seek and can retain higher-paying positions. Second, employers who minimize employee pay tend to lose their best workers to other organizations that are willing to pay more.

Consequently, when less-skilled workers are attracted to an organization, and when the best of an organization's workers ultimately seek employment elsewhere, product quality and customer service levels inevitably are below average. This, in turn, results in reduced sales and smaller company profits. An optimum compensation program attracts high-quality workers and allows the company to maximize profitability.

When assessing compensation packages, foodservice operators first consider the salaries, wages paid per hour, and/or tips received during a worker's average shift. However, operators designing competitive compensation packages must consider much more than the amount of money paid to their workers. This is so because an effective compensation program consists of important **extrinsic rewards** and **intrinsic rewards**.

For most employees, both extrinsic and intrinsic rewards are important. As a result, foodservice operators must carefully consider both types of rewards when developing their total compensation programs. Figure 10.1 lists some of the most common financial and nonfinancial extrinsic rewards used in the foodservice industry. Figure 10.2 lists some of the most common intrinsic rewards.

Key Term

Extrinsic rewards: Financial and non-financial compensation granted to a worker by others (usually the employer).

Key Term

Intrinsic rewards: Self-initiated compensation (e.g., pride in one's work, a sense of professional accomplishment, and/or enjoying being part of a work team).

Extrinsic Rewards	
Financial	**Nonfinancial**
Salaries	Preferred office space or workstation
Hourly pay	Preferred personal computer
Cost of living adjustments ("Colas")	Preferred kitchen tools
Tips	Preferred meal privileges
Commissions	Designated parking place
Bonuses	Business cards
Merit pay	Special dress code
Incentive pay	Special designations on uniforms
Profit sharing	Impressive titles
Paid leave	
Mandatory benefits	
Voluntary benefits	

Figure 10.1 Extrinsic Employee Rewards

Intrinsic and extrinsic rewards are essentially polar opposites. One reward is internally based, while the other is externally based. For example, an intrinsic reward is the feeling of satisfaction a team of foodservice workers may experience when they redesign an operation's menu. An extrinsic reward is the monetary bonus each team member might receive for their contributions to the menu redesign project.

Not all team members react in the same manner to the rewards offered by their employers. For some workers, intrinsic rewards are critically important. For others, financial rewards may be most important, and for still others, status and nonfinancial extrinsic rewards may be what they like most about their compensation.

Intrinsic Rewards
Participation in job design
Participation in decision-making
Greater job freedom
More interesting work
Pride in one's work
Opportunities for personal growth
Opportunities for professional growth
More job security
Empowerment

Figure 10.2 Intrinsic Employee Rewards

All paid employees exchange work for rewards. While not all employees earn the same amount of money, nearly all employees view the amount they receive as a real indication of their value from the perspectives of their managers. Therefore, an employee who discovers that a co-worker makes even as little as 5 or 10 cents more per hour than they do may become quite upset.

The most important thing for foodservice operators to understand about the way team members view compensation is that the amount of money earned significantly enhances or detracts from their employees' own feelings of status and self-worth. Therefore, an equitable compensation program that considers pay and other employee rewards is critical.

What Would You Do? 10.1

"It's not fair," said JeAnna, "we do the same job, I work just as hard as she does, and now I find out she's making $1.50 an hour more than I am!"

JeAnna was talking to Konstantin Papadopoulos, the owner/operator of the Athena Grill, a "taverna" (a casual operation that served grilled meats and fried seafood along with popular Greek side dishes).

JeAnna, who had worked at the grill for six months, had asked to speak to Constantine because she had learned that Nefeli, another staff member at the grill, was doing the same job as JeAnna, but was making a different hourly rate of pay.

"Nefeli has been with us for five years," said Konstantin.

"That doesn't make it right!" replied JeAnna, "we should all get equal pay for equal work!"

Assume you were Constantine. One popular and effective way of retaining high-quality workers is to reward them financially for their longevity. The result, however, is that frequently two different workers who are doing the same job will receive different hourly rates of pay. How would you explain the "fairness" of that compensation policy to JeAnna?

Compensation of Team Members

The goal of any effective **compensation management** program should be to attract, motivate, and retain competent employees. To do this, the program must be perceived by employees as fair and equitable.

While there are a number of ways foodservice operators can assess their overall compensation

Key Term

Compensation management (program): The process of administering a foodservice operation's extrinsic and intrinsic reward systems.

management efforts, one good way to do so is to consider the following key program segments:

✓ Legal Aspects of Compensation
✓ Payment of Salaries
✓ Payment of Hourly Wages
✓ Compensation of Tipped Employees
✓ Benefit Payments

Legal Aspects of Compensation

Regardless of how foodservice operators plan and administer their compensation management programs, all aspects of the program must follow applicable federal, state, and local employment laws (see Chapter 2).

While all compensation programs must comply with federal and state laws, even county or city laws must be known and applied if an effective and legally compliant compensation program is to be developed. Fines for noncompliance with employee wage payment requirements can be significant and they also can, if widely publicized, significantly damage the reputation of a foodservice operation. Therefore, foodservice operators must begin their compensation program development with a thorough review of applicable laws and ordinances in the areas in which they operate their businesses.

Payment of Salaries

In the foodservice industry, salaried employees are most often managers and supervisors. Salaried employees are more accurately described as **exempt employees**. The reason: their duties, responsibilities, and levels of decisions make them "exempt" from the overtime provisions of the U.S. federal government's Fair Labor Standards Act (FLSA).

Key Term

Exempt employee: An employee who is not subject to the minimum wage or overtime provisions of the Fair Labor Standards Act (FLSA).

Exempt employees do not receive overtime when they work more than 40 hours per week. They are also expected by most foodservice operators to work the number of hours needed to adequately perform their jobs.

Job titles do not determine exempt status. Rather, specific job duties and salary must meet all requirements of the Department of Labor's regulations to qualify for exempt status. In the foodservice industry, most exempt jobs fall into either executive, administrative, or supervisor categories.

In general, to qualify for the executive employee exemption, all the following tests must be met:

✓ The employee must be compensated with a minimum salary established by federal law.

✓ The employee's primary duty must be managing the operation or managing a customarily recognized department or subdivision of the operation.
✓ The employee must customarily and regularly direct the work of at least two or more other full-time employees or their equivalent.
✓ The employee must have the authority to hire or fire other employees. Alternatively, employees' suggestions and recommendations as to the hiring, firing, advancement, promotion, or any other change of status of other employees must be given significant weight.

Find Out More

Some foodservice operators believe that making an employee a salaried worker rather than paying that worker an hourly rate is a good way to save money because no overtime pay must be provided to the worker. The difference between exempt and non-exempt employees is clear: exempt employees are exempt from overtime pay. However, the definition of exactly who qualifies as an exempt employee varies by state.

Employers must accurately classify employees as exempt or non-exempt. Misclassification can result in heavy fines and, without a firm grasp on the distinctions, an employer cannot accurately forecast future payroll costs. Even experienced foodservice operators can sometimes make an error in their understanding about the details of what makes an employee exempt or non-exempt.

To gain a better understanding of who does and does not qualify as an exempt worker in a specific state, enter "exempt worker requirements in (name of state)" in your favorite search engine and review the results.

From a compensation package perspective, the most important thing for foodservice operators to remember is that salaried workers will view the fairness of their compensation based on two distinct factors. One is the total amount of salary paid. Clearly, higher salaries are considered more desirable by workers than lower salaries.

Just as important, however, is the number of hours the salaried employee is expected to work. Foodservice operators must resist the temptation to overwork salaried staff members. One reason is because these workers may leave an organization to accept a different position that offers a higher salary, or even a *lower* salary, if the number of hours required to complete the job is significantly less.

The number of hours a salaried worker *should* work is a subjective question. However, operators must be realistic and understand they should view salaries paid from the perspective of their salaried employees, as well as their own.

Payment of Hourly Wages

Developing compensation packages for hourly paid employees involves the management of employees' expectations and perceptions. An effective hourly pay compensation program typically includes:

1) *Categorization of jobs*—Not all employees do the same work and the result is that employees' hourly pay differences exist. Most employees will readily accept this rationale as the reason for pay variations. It is easy for most employees to understand, for example, that an experienced cook in a foodservice operation would likely make more per hour than a dishwasher in the same operation. When employees understand real differences in job responsibilities, they can better understand the reasons for differences in hourly pay.

   Within job categories, it is also reasonable to develop pay ranges (see Chapter 5). For example, cooks in a restaurant may be classified as trainee, intermediate, senior, and the like to designate different experience or skill levels. Each classification would, under this system, have its own pay range. Employees can also be made aware of the skills, experience, or longevity needed to advance to higher levels. Also, they can also be informed about opportunities to help them become trained or eligible for these higher-paid positions.

2) *Management of internal pay equity*—Most foodservice operators would agree that managing internal pay equity is important. Employees are very likely to know the hourly pay or salaries of their co-workers. Internal equity is best achieved by paying people within the pay range established for their jobs and by varying pay for objectively identifiable measures. These include job performance, full- or part-time status, shifts worked, assignments completed, and other measurable factors.

3) *Linkage of pay to job performance*—Most foodservice operators and their employees probably agree that workers who perform their jobs better should receive a larger hourly pay rate (and larger pay increases) than those who do not perform as well. Chapter 11 of this book addresses how operators can evaluate their employees' performance and contributions to the success of their businesses, and then use that information to adjust hourly pay rates.

It is important to recognize that, as is the case with salaries, hourly paid employees will evaluate their total compensation package based on both the hourly rates they are paid and the number of hours they are scheduled to work.

Compensation of Tipped Employees

Foodservice operations are somewhat unique because they allow employee tips as part of their wage payments when satisfying Federal Labor Standards Act (FSLA) worker pay requirements. As a result, one challenge faced by foodservice operators relates to the accounting procedures required to compensate tipped employees fairly and legally and to maintain proper records of doing so.

Managing compensation packages for tipped employees is a challenge in nearly all segments of the foodservice industry. Tips actually received by tipped employees may be counted (credited) as wages for purposes of the FLSA, but the employer must still pay a minimum amount established by law.

If operators elect to use the tip credit provision of the law, they must:

1) inform each tipped employee about the tip credit allowance (including the amount to be credited) before the credit is utilized;
2) be able to show that the employee receives at least the minimum wage when direct wages and the tip credit allowance are combined; and
3) allow tipped employees to retain all tips unless they participate in a valid multi-employee tip-distribution arrangement.

If an employee's tips combined with the employer's direct wages do not equal the federal minimum hourly wage, the employer must make up the difference. Current law forbids any arrangement between the employer and a tipped employee in which any part of the tip received becomes the property of the employer; because a tip is the sole property of the tipped employee.

When an employer does not strictly observe the tip credit provisions issued by the FLSA, no tip credit may be claimed. Then employees are entitled to receive the full cash minimum wage plus all tips they have received.

When tips are charged on a credit card, and the employer pays the credit card company a percentage on each sale, the employer may pay the employee the tip *minus* that percentage. However, the charge on the tip may not reduce the employee's wage below the required minimum wage. The tip amounts due from payment cards must be paid no later than the employee's regular payday and cannot be held while the employer is awaiting reimbursement from the payment card company.

In most foodservice operations, a modern point-of-sale (POS) system is used to record the credit card and cash tips guests intend to give to their servers. Increasingly, however, some foodservice operations use a **tip distribution program** to manage the compensation of tipped employees.

For example, assume a tip distribution program is in place, and a guest gives a server a $40.00 tip on a $200.00 guest check that

Key Term

Tip distribution program: A system of tip payment that allows an operation to distribute a customer's tip from an employee who actually received it to others who also provided service to the customer.

included $150.00 of food and $50.00 of alcoholic beverages. The tip program assigns a portion of the tip to the guest's food server and another portion to the bartender(s) who prepared the guest's drinks.

To accurately record and pay out tips, tip distribution programs are typically needed when an operation has either a **tip sharing** or **tip pooling** system in place.

The best tip distribution programs are interfaced directly to a restaurant's payroll accounting system. The reason: if an employee "customarily and regularly" makes more than $30 per month in tips, then current federal law requires that the employee is considered a tipped employee for minimum wage and overtime pay purposes. Many states also have special laws for tipped employees, and some have different standards for qualification as a tipped employee rather than the federal standard.

Key Term

Tip sharing: A tip system that takes the tips given to one group of employees and provides a portion of them to another group of employees. Used, for example, when server tips are shared with those who bus the server's tables.

Key Term

Tip pooling: A tip system that takes the tips given to individual employees in a group and shares them equally with all other members of the group. Used for example, when bartender tips given to an individual bartender are shared equally with all bartenders working the same shift.

Technology at Work

Tip distribution software, also referred to as tip pooling or gratuity management software, automates the process of paying tips to tipped workers at the end of their shifts and reduces the need for physical cash payouts.

These tools eliminate hours of manual labor by tracking the number of hours worked and automatically distributing tips to workers' bank accounts. These tools can also reduce payroll burdens and instances of tip disparity due to errors or theft. They also expedite paying out tips to employees by connecting securely to their bank accounts or debit cards.

Additionally, most tip distribution software allows operators to manage tips across multiple locations and utilize proper reporting features to reduce risks of non-compliance in tip payments.

To examine some of the offerings of companies that have developed software to assist in tip payment and proper recording, enter "tip distribution software in restaurants" in your favorite search engine and review the results.

Most tipped employees recognize the guest, not the foodservice operator, largely determines the level of tips they earn. However, while the quality of service provided to guests impacts tips, the specific days and hours employees work are also a factor in the amount of tips earned.

For this reason, as foodservice operators create compensation packages for tipped employees, they should recognize their responsibilities to schedule workers in a fair and equitable way. Inequitable favoritism in shift assignment will always be recognized (and resented!) by tipped employees.

Benefit Payments

Benefit costs for employees are incurred in every foodservice operation and, as a result, they become part of each worker's compensation package. One good way to consider the benefit components of a compensation package is to examine them from the perspective of being either a **mandatory benefit** or a **voluntary benefit**.

Under current federal and state laws mandatory employee benefits include:

✓ Social Security and Medicare
✓ Unemployment insurance
✓ Workers' compensation insurance
✓ Family and Medical Leave Act (FMLA) protections

Key Term

Mandatory benefit: An employee benefit that must, by law, be paid by an employer.

Key Term

Voluntary benefit: An employee benefit paid at the discretion of an employer in efforts to attract and keep the best possible employees.

Social security and medicare. The Federal Insurance Contributions Act (FICA) is a federal payroll (employment) tax used to fund Social Security and Medicare programs. Current law states that both employees and employers are required to contribute to these funds.

Unemployment insurance. Employers are required to contribute to unemployment insurance through payroll taxes at both the state and the federal level to assist workers who lose their jobs.

Unemployment insurance protects both part- and full-time employees who meet certain requirements and are separated from a company by providing some income for a limited period. Since unemployment insurance is managed by the individual states, the cost of this insurance and amount required for each employer varies from state-to-state. However, all states in the United States have some minimum requirements for unemployment insurance, and all employers must participate in their state's program and carry at least the minimum required amount of coverage.

Workers' compensation insurance. Workers' compensation insurance provides financial support to people unable to work as a result of a workplace injury or illness. If an employee experiences an injury or illness as the result of their regular on-the-job duties, states mandate that the employer should be responsible for covering medical bills as well as a limited amount of income for the employee during the recovery period. Worker compensation rates to be paid by an employer are set by the various states.

Family and medical leave act protections. The Family and Medical Leave Act (FMLA) (see Chapter 2) is the federal law that entitles eligible employees of covered employers to take unpaid, job-protected leaves for specified family and medical reasons. A covered employer is a private-sector employer with 50 or more employees and all public employers. The FMLA provides eligible employees with up to 12 weeks of job-protected, unpaid leave during a 12-month period for qualifying family and medical reasons. Qualifying reasons include the birth of a child, dealing with a serious or chronic personal illness, or caring for an immediate family member with a serious or chronic illness.

Note: In addition to benefits under the FMLA, some states and local jurisdictions require paid/unpaid family leave and/or paid/unpaid sick and **safe leave**. Foodservice operators should review their obligations under applicable state and local laws.

> **Key Term**
>
> **Safe leave:** Absence from work when an employee or employee's family member has been the victim of domestic abuse or violence.

While the costs of providing mandatory benefits are real, from a compensation package perspective all foodservice operators must offer current and potential team members these same benefits. Therefore, mandatory benefits provide operators with little ability to differentiate themselves among employers when seeking to attract and retain high-quality workers. The provision of voluntary benefits, however, does provide operators with the ability to make their compensation packages more attractive.

Voluntary benefits offered by an employer can make a significant difference in a potential worker's decision to join and stay with a specific foodservice operation. Voluntary benefits offered by foodservice operators can include:

✓ Health insurance
✓ Retirement contributions
✓ Paid time off
✓ Bonuses
✓ Free meals

Health insurance. While mandated for large employers, the Patient Protection and Affordable Care Act (see Chapter 2), provides that foodservice operations in the United States with less than 50 full-time or **full-time equivalent** employees are not required to offer health insurance for employees. In the restaurant industry, approximately 35% of operations offer access to medical insurance. This is well under the national average of 69%.[1] In many cases, when health insurance coverage is offered, it is offered only to salaried employees.

Regardless of whether a foodservice operator must offer health insurance, offering a group health plan can help increase employee retention and loyalty. The reason: by showing employees that the operation cares about their long-term well-being. In addition, providing a work team with access to the medical support it requires may lead to fewer staff absences and sick days.

One example of a foodservice operator going above and beyond in worker health-related benefits is In-N-Out Burger. This chain of quick service restaurants offers full-time and part-time employees and their dependents the option of enrolling in dental, vision, voluntary life, and accidental death and dismemberment (AD&D) insurance. Note: These programs are in addition to vacation and sick benefits and free meals.[2] In fact, surveys of foodservice workers consistently reveal that access to affordable health care is their number one benefit preference.

Retirement conditions. All workers recognize they will, at some point, end their working lives. Planning for retirement will likely be of little interest to part-time high school students working in a foodservice operation. However, full-time workers and those who choose the industry as a profession will be very concerned about their retirement savings. For example, one voluntary benefit desired by many foodservice workers is the ability to contribute to their own **401-k**. These workers are also especially attracted to

Key Term

Full-time equivalent (employee): A unit of measurement used to calculate the number of full-time hours worked by all employees in a business.

For example, if a business considers 40 hours to be a full-time workweek, then an employee working 40 hours per week would have an FTE of 1.0. In contrast, a part-time employee working only 20 hours per week would have an FTE of 0.5 — which shows that the hours they worked are equivalent to one half of a full-time employee.

Key Term

401-K: A retirement program where an employee can make contributions from their paycheck either before or after-tax, depending on the options offered in the plan. If they desire to do so, employers can also make contributions to a worker's 401-k account.

---

1 https://mployeradvisor.com/state-benefit-guides/employee-benefits-summary-for-the-restaurants-and-bars-industry. Retrieved October 26, 2023.
2 https://totalfood.com/employee-benefits-key-to-attracting-and-retaining-employees/. Retrieved June 25, 2023.

employers who agree to match all, or part, of the employee's contribution to the plan.

Paid time off. Approximately 20 states have mandatory paid sick leave laws. So, while not required by law in all states, a voluntary **paid time off (PTO) program** is valuable because it can help a foodservice operation's team members feel appreciated and well-treated.

Key Term

Paid time off (PTO) program: An employer-provided benefit in which an employee is allotted an amount of paid time that can be used for vacation, sickness, or personal time at their discretion.

PTO programs designed by a foodservice operator can range from very modest to quite extensive. Operators can also establish "time on the job" requirements to qualify for the program. For example, a foodservice operator may offer employees one week's paid vacation after an employee has worked for 12 months, and two weeks' paid vacation if the employee has worked in the operation for five years or more.

Other components of a PTO program can include payment for sick days and holidays when an operation is closed. It is important to note that the presence of a PTO program (even a modest one!) will make an operation more appealing to most foodservice workers than will a competitor's operation that offers no PTO program.

Bonuses. Historically, bonuses in the foodservice industry have been paid to salaried foodservice managers and supervisors who achieve specific perfor-mance targets and goals. A bonus of this type can often spur supervisory staff to do their best, but bonuses, even modest ones, can also motivate and help to retain high quality hourly paid staff member.

Note: Foodservice operators can be creative in this area. For example, a mod-est bonus might be paid if a foodservice employee does not miss an assigned shift for 90 days in a row. Some foodservice owners implement profit sharing as part of their bonus programs. This tactic not only rewards long-term employ-ees, but it can help motivate employees to do their best work every day of the year.

Free meals. Providing free meals for restaurant employees is a common volun-tary benefit. Some employers restrict the specific menu items that can be con-sumed by workers. Other employers charge workers a percentage of the menu price and then allow them to select the items they wish.

Regardless of the approach taken, hungry employees will appreciate a nice break for a meal. Also, they can try different menu items, and this opportunity helps them to better understand how to sell the items (if they are servers) or prepare the items (if they are production personnel).

Some potential voluntary benefits are unique to a foodservice operation. For example, if a foodservice operates in a hotel, reduced guest rooms for visiting family or friends may be offered. Those foodservice operators working in popular attractions such as amusement parks, zoos, museums, sports venues, and theaters may also be able to provide employees with reduced cost or no cost tickets for admittance.

A concern voiced by some foodservice business owners and operators relates to the cost of providing voluntary benefits. In many cases, these professionals would like to offer workers more benefits because they know benefits are important to workers but feel the costs of doing so are too great. In many cases, this attitude is the result of a misunderstanding of foodservice pricing. To better understand why this is the case, recall from earlier in this chapter that the formula used to calculate a labor cost percentage is:

$$\frac{\text{Total cost of labor}}{\text{Total revenue (sales) generated}} = \text{Labor Cost \%}$$

As the top line (numerator) in the formula increases labor cost % will, of course, increase as well. But it is important to recognize that as the lower line (denominator) increases labor cost % will decline.

The amount of total revenue generated by a foodservice operation depends on the prices it charges for the products and services it sells. Foodservice operators may believe that, if they raise prices to generate revenue to contribute to the cost of labor, their total revenue will decline, because competitors offer lower prices.

The prices charged by an operation's competitors can be important, but this factor is sometimes too closely monitored by some foodservice operators. It may seem to some operators that their average guest is only concerned with low prices. However, small price variations generally make little difference in the buying behavior of the average guest.

For example, if a group of young professionals goes out for pizza and beer after work, the major determinant is not likely whether the selling price for a small pizza is $17.95 in one operation or $19.95 in another facility. Other factors including quality, location, and parking availability may become more important.

The selling prices of potential competitors are of concern when establishing selling prices, but experienced operators understand that a specific operation can always sell a product of lower quality for a lesser price. While competitors' prices can help an operator arrive at their property's own selling prices, it should not be the only determining factor.

The most successful foodservice operators focus on building guest value in their own operations and not in attempting to mimic their competitors' efforts. Even though operators may believe their guests only want low prices, remember that consumers often associate higher prices with higher quality products and, therefore, products that provide a better price/value relationship. It is interesting to note foodservice operations that have been successful over the long term, and they often use superior service levels rather than lower prices to grow their businesses.

Foodservice operators should have confidence in the products and services they provide. When they do, they can charge prices that allow them to develop compensation packages that attract and retain the best work teams possible and still make significant profits. The importance of pricing properly was well put by Warren Buffett, consistently ranked in the top ten of the world's most wealthy individuals when he stated:

> *The single most important decision in evaluating a business is pricing power. If you've got the power to raise prices without losing business to a competitor, you've got a very good business. And if you have to have a prayer session before raising the price by 10 percent, then you've got a terrible business.*[3]

Sometimes foodservice operators hurt themselves and the entire industry when they seek to pay extremely low wages, and, as a result, they find that they cannot attract high-quality staff members.

A final word about voluntary benefits is that it can be expensive for a foodservice operation to provide attractive benefits to employees. However, it can be more expensive not to provide them, and to lose worker talent as a result. Employee turnover is expensive, especially when an operation loses key staff with valuable experience. Many workers choose and keep jobs with voluntary benefits in mind, especially in a hiring landscape where actual wages are relatively similar across potential employers.

Voluntary benefits offered are particularly important for employee retention. Health care benefits are cited by most foodservice workers to be the single benefit that is most effective in encouraging them to remain with their jobs. Retirement benefits and paid time off are also high on the list of benefits that encourage employees to stay.

Scheduling Team Members

Foodservice workers are concerned about their pay and benefit packages. However, many hourly workers and some salaried workers state that their assigned work schedules are of equal concern. For that reason, foodservice operators must take great care as they prepare and distribute their **master schedules**.

Key Term

Master schedule: A listing of all employee shifts including specific start and end times for each shift during a calendar week.

---

3 https://www.raybusinessadvisors.com/warren-buffett-on-pricing-power-and-your-idiot-nephew/. Retrieved June 25, 2023.

While master work schedules can be produced for any period, they are produced and distributed weekly in most foodservice operations. The production of master schedules can range in simplicity from the use of methods like pen and paper, to printed spreadsheets, and automated techniques utilizing employee scheduling software.

Experienced foodservice operators recognize that a thoughtfully prepared master work schedule improves employee satisfaction, reduces absenteeism, and increases productivity. The process of producing a master schedule can be complicated because of multiple variables including forecasted sales volume, employee availability, and staff time-off requests (including PTO) considerations. As well, seasonal requirements can also affect the development of a high-quality master work schedule.

While the actual process used to develop a master schedule varies based on the size and specific staffing needs of a foodservice operation, all operators responsible for work schedule preparation must understand two key concepts:

1) The legal aspects of scheduling
2) The schedule development and distribution process

Legal Aspects of Scheduling

Foodservice operators have great freedom in developing their master schedules, but that freedom is not absolute. To illustrate, many foodservice operators employ young workers. The FLSA generally sets 14 years old as the minimum age for employment, and it limits the number of hours worked by minors under the age of 16. However, specific laws related to child labor vary by state.

To cite just one example of state law impacting scheduled development, Wisconsin has no limit during non-school week on daily hours or night work for 16- and 17-year-olds. However, persons of this age must be paid time and one-half for working more than 10 hours per day or 40 hours per week, whichever is greater. Also, 8 hours' rest is required between the end of work and start of work the next day, and any work between 12:30 AM and 5:00 AM must be directly supervised by an adult.

In a similar manner, many state and local scheduling laws require penalty payments to employees who are scheduled to work back-to-back shifts without sufficient off-duty hours in-between the shifts. Modifications of work schedules can also put operators in violation of the law or impact wage payments.

In California, for example, when an employee shows up to work but works less than half of their assigned shift and is sent home early, the employer is responsible for "reporting time pay." This penalty is one-half of the worker's regularly scheduled shift, but not less than two hours and not more than four hours. For example, assume a worker was scheduled for eight hours but was sent home early after working only three hours (less than half of the scheduled shift). In this example, the worker would be owed reporting time pay, and the employer must pay for half of the employee's scheduled shift (four hours).

Under New York City's Fair Workweek Law, covered employers must post schedules 72 hours in advance, and changes thereafter are prohibited. Exemptions apply if the employer cannot operate due to certain emergency conditions such as natural disasters, failures of public utilities, or shutdowns of public transportation.

The above examples illustrate that foodservice operators must be knowledgeable about all federal, state, or local laws that impact how work schedules must be prepared and implemented.

Find Out More

While federal law dictates several aspects of employee scheduling and payment, many states and local governments have passed additional wage-related laws that directly impact how employees may be scheduled and paid.

It is important to recognize that modifications to these laws are made regularly. Therefore, all foodservice operators must stay up-to-date with state and local legislation that impacts their scheduling efforts.

To learn more about how a specific state's laws impact the scheduling of foodservice workers, enter "state laws affecting employee scheduling and payment in (insert name of state)" into your favorite search engine and review the results.

Schedule Development and Distribution

The development and distribution of the master work schedule is one of a foodservice operator's most important tasks, and it must be done well. This can be a four-step process as shown in Figure 10.3.

Step 1: Forecast sales volume

The first step to develop a master work schedule is to create an accurate sales forecast. In all but the smallest of foodservice operations, the number of staff needed on a specific day will most often depend on an operator's best estimate of the number of guests that will be served on each day. For example, in many operations weekends

Figure 10.3 Four Step Master Work Schedule Development Process

will generate more volume than weekdays. In other operations, the opposite may be the case.

The cost of hourly paid labor is, in large part, a **variable cost**. Therefore, the process of determining the total scheduled employee work hours must correlate closely to sales volume if an operator seeks to maintain targeted or desired labor costs and labor cost percentages.

Key Term

Variable cost: An expense that generally increases as sales volume increases and decreases as sales volume decreases.

If too few workers are scheduled based on anticipated volume, guest service quality will, in many cases, suffer. Alternatively, if too many workers are scheduled, labor costs will be excessive.

While hourly paid labor can be considered a variable cost, foodservice operators must also understand that hourly paid labor can also be viewed as a **step cost**.

Key Term

Step cost: A cost that increases in a non-linear fashion as activity or sales volume increases.

A step cost increases as a range of activity increases or as a capacity limit is reached. Instead of increasing in a linear fashion like many variable costs, step cost increases look more like a staircase (the reason for the name "step" costs).

It is easy to understand step costs. Consider the foodservice operator who determines that one well-trained server can effectively provide excellent service for a range of 1–30 guests. If, however, 40 guests are anticipated, a second server must be scheduled. However, the operator does not need a "full" server, and a 1/3 server would be sufficient because only 10 additional guests are anticipated. As shown in Figure 10.4, however, each additional server added increases the operation's costs in a non-liner (step-like) fashion.

Step costs may affect both hourly and salaried workers. For example, in a large foodservice operation one salaried banquet supervisor may be able to direct the work activities of eight servers. When nine or more servers are scheduled to work, additional banquet supervisors must be added. These additional salaried staff members will also increase the operation's "day of banquet" payroll costs in a step-like fashion.

Understanding the relationship between anticipated volume and worker productivity (see Chapter12) is the key to recognizing the number of hourly employees needed on a specific day. The schedule planner must be able to anticipate the day's number of guests to be served and/or sales dollars to be generated prior to creating the master schedule.

Step 2: Develop the schedule

As they develop master schedules, foodservice operators must consider anticipated volume and numerous limitations on staff assignments. Issues such as

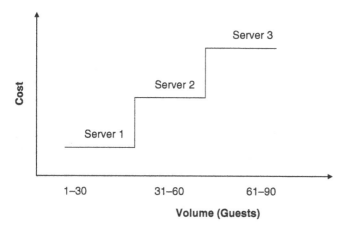

Figure 10.4 Sales Volume and Step Costs

minimizing overtime hours, staff shift availability and requests, legal limitations, and other factors can all impact developing the master schedule.

A properly prepared master schedule will list each employee assigned to work and provide specific information including:

✓ Employee's first and last (full) name
✓ Assigned work day
✓ Assigned work date
✓ Start of shift time
✓ End of shift time
✓ Specific work assignment/location (for workers who may fill multiple roles)
✓ Lunch/other breaks (if applicable)

Step 3: Distribute the Schedule

Methods used by foodservice operators to inform their staff about the anticipated schedule vary. Sometimes the schedule is posted in a central on-premises location such as the bulletin board outside a manager's office and/or in an employee break room.

Increasingly, however, foodservice operators find that posting work schedules online make the most sense. When employees are given online access to their work schedules, they can check them at any time and from any location and are not required to be on property to learn this information.

While there are no federal regulations mandating the amount of time in advance that a schedule must be distributed, several states and local governments have enacted **predictive scheduling legislation** (and some states have specifically prohibited it!) that can impact schedule distribution.

Key Term

Predictive scheduling legislation: Laws addressing the minimum amount of advance notice of work schedules that employers must provide to their employees.

Proponents of predictive scheduling legislation observe that these schedules benefit part-time employees with fluctuating schedules. Note: They also enable full-time employees to better estimate their cash flow and maintain a healthy work/life balance. Predictive scheduling legislation may require employers to post work schedules between 72 hours and two weeks in advance.

Step 4: Modify the schedule as needed

In many foodservice operations an hourly employee's specific work schedule is often difficult to predict. As a result, modifications of the master schedule are common. Changes in revenue forecasts may cause a foodservice operator to require more or fewer staff hours than originally anticipated. Employees who call in sick or no-show for their assigned shifts also impact the master work schedule.

In the past, there has been little or no regulation requiring employers to provide advanced notice of schedule modification. However, in some locations predictive scheduling legislation place limitations regarding how and when employers may modify a posted schedule. Most predictive scheduling laws require employers to provide additional pay to employees if the employer changes the schedule.

Technology at Work

Foodservice operators and others who create weekly master work schedules for more than only a few employees often find the process to be challenging and time consuming.

Essentially, the development of an effective schedule balances the staffing needs of the operation with employee desires. The process becomes complicated when various employees have different requests, and their availability along with business volume levels (and the need for staff) fluctuates.

Fortunately, a significant number of companies offer staff scheduling software designed specifically for the foodservice industry. The best of such programs:

1) Analyze historical sales data and incorporate external factors such as weather reports and local events when determining staffing needs.
2) Automatically incorporate PTO and non-paid time off requests of staff.
3) Provide an easy way for staff members to view work schedules on-line.
4) Allow for real-time schedule up-dates as needed.
5) Forecast total labor costs in dollars and labor cost percentages based on the established schedule and revenue forecasts.

To see examples of software that create work schedules meeting operational needs and work team preferences, enter "staffing software for restaurants" in your favorite search engine and review the results.

This and previous chapters addressed how foodservice operators legally recruit, select, orient, train, compensate, and schedule work teams to help ensure the best possible results. Sometimes, however, problems with the quality or quantity of an employee's work arise. Also, an employee's work may be acceptable, but difficulties about compliance with workplace rules and procedures are discovered.

Managers must fairly evaluate the quality of their employees' efforts and, if necessary, correct and improve their performance. Doing these things in a legal and effective manner is extremely important and are the topics of the next chapter.

What Would You Do? 10.2

"I'm sorry, but I need everybody on payroll to be working on Sunday. It's not personal. We're going to be swamped with customers, and I need everybody to work all day," said Ricky, the operator of the Lone Star Barbecue House.

Ricky was talking to Lea Dobson, one of Lone Star's best and most experienced servers.

The Lone Star was located near City Community College, which was holding its spring graduation on Sunday. The graduation had originally been scheduled for Saturday, but the graduation ceremony was planned to occur outdoors, and Saturday's forecast (rain all day) prompted the college to move the graduation ceremony to Sunday.

"My son is graduating on Sunday," replied Lea, "he's the first one in our family ever to graduate from college, and I can't miss the graduation. That's why I put in my request months ago to have Saturday off and to work Sunday. But now that the graduation has been moved to Sunday, I can't come in."

"Yes," replied Ricky "but it was only a request, and now that the graduation has been moved to Sunday, I can't grant your request."

"Then I guess I'll be a no-show on Sunday," replied Lea, "if that means you should fire me then do what you have to do, but I will not be here Sunday because I am going to see my son graduate!"

Assume you were Ricky. Do you think it would be in the best interest of your business to terminate Lea if she would "no-show" on Sunday? What message would it likely send to your remaining staff members if she were terminated? What message would it send to your employees if she were not terminated? What would you do about Lea's request?

Key Terms

Compensation	Exempt employee	401-k
Labor cost percentage	Tip distribution program	Paid time off
Salaried employee	Tip sharing	(PTO) program
Compensation package	Tip pooling	Master schedule
Extrinsic rewards	Mandatory benefit	Variable cost
Intrinsic rewards	Voluntary benefit	Step cost
Compensation	Safe leave	Predictive scheduling
management	Full-time equivalent	legislation
(program)	(employee)	

Operator's 10-Point Tactics for Success Checklist

Evaluate your need for, and the current status of, each of the following operational tactics. For those tactics you think are important, but not yet in place, develop an action plan for its implementation including who will be responsible for the tactic's completion and the target date by which it should be completed.

				If Not Done	
Tactic	**Don't Agree (Not Done)**	**Agree (Done)**	**Agree (Not Done)**	**Who Is Responsible?**	**Target Completion Date**
1) Operator recognizes the importance of fair compensation to the maintenance of a productive work team.	——	——	——		
2) Operator can summarize the main components of a business owner's view of compensation.	——	——	——		
3) Operator can summarize the main components of a foodservice operator's view of compensation.	——	——	——		
4) Operator can summarize the main components of a staff member's view of compensation.	——	——	——		

				If Not Done	
Tactic	**Don't Agree (Not Done)**	**Agree (Done)**	**Agree (Not Done)**	**Who Is Responsible?**	**Target Completion Date**
5) Operator recognizes the legal aspects related to the development of employee compensation packages.	——	——	——		
6) Operator can state the main differences between paying a staff member a salary compared to an hourly wage.	——	——	——		
7) Operator has reviewed the special payment circumstances associated with compensating tipped employees.	——	——	——		
8) Operator recognizes the importance of voluntary and involuntary benefits as key components of an effective employee compensation package.	——	——	——		
9) Operator recognizes the importance of knowing federal, state, and local laws impacting the development of employee work schedules.	——	——	——		
10) Operator can list the four steps required in the development and distribution of employee master work schedules.	——	——	——		

11

Appraising and Managing Staff

What You Will Learn

1) The Importance of Staff Appraisal
2) How to Create a Performance Appraisal Program
3) How to Implement a Progressive Discipline Program

Operator's Brief

In this chapter, you will learn that to develop the most efficient work teams possible, your efforts to improve the quality of your staff must be on-going. Most foodservice workers want to do a good job and, as an operator, you will be able to help them improve with formal performance management and performance appraisal programs.

A performance management program is a systematic process by which operators can help employees improve their abilities to achieve goals. The program consists of an objective and comprehensive evaluation of employee effectiveness.

Operators who have determined that a performance appraisal program should be instituted in their businesses can choose from several popular approaches:

✓ Absolute Standards
✓ Relative Standards
✓ Targeted Outcomes

In this chapter, you will learn about the advantages and potential drawbacks of each of the above popular alternatives.

Inevitably, you will find that some staff members must be disciplined to improve their performance. The use of a progressive discipline program is recommended, and, in this chapter, you will learn about the four steps that are commonly used to implement and manage a proper program:

✓ Documented Oral Warning
✓ Written Warning
✓ Suspension
✓ Dismissal

In some cases, employee separations from your business might be inevitable. They may, in some cases, be voluntarily initiated by staff members, and, in other cases, you must initiate the separation. Both types of separations are addressed in this chapter.

Finally, in some instances you will find that employee exit interviews can be of value in improving your operation. This chapter discusses the types of questions that can be addressed in an exit interview that may help you to reduce future staff turnover.

CHAPTER OUTLINE

Performance Appraisal Programs
    The Importance of Regular Performance Appraisal
    The Characteristics of Effective Performance Appraisal
    Benefits of Performance Appraisal Programs
Popular Performance Appraisal Methods
    Absolute Standards
    Relative Standards
    Targeted Outcomes
Progressive Discipline Programs
    Documented Oral Warning
    Written Warning
    Suspension
    Dismissal
Staff Separation
    Voluntary Separation
    Involuntary Separation
    Exit Interviews

Performance Appraisal Programs

Previous chapters have addressed how foodservice operators recruit, select, train, and fairly compensate their employees to help ensure they develop and maintain the best possible staff. These are continual processes because workers' jobs continually evolve as guests' needs and desires change, and as new menu items and/or new work methods are implemented.

Efforts to continually improve a work team's job performance must be ongoing and, in most cases, these efforts will be successful. Sometimes, however, problems with the quality or quantity of a team member's work arise. In other cases, an employee's work may be acceptable, but difficulties about compliance with workplace rules and/or procedures arise. When this happens, foodservice operators must fairly evaluate the quality of their employees' efforts and, when necessary, correct and improve their performance.

Most staff members want to do a good job and effective foodservice operators can often help them do so. Knowing how to objectively evaluate and improve worker performance and, when necessary, properly terminate employees are important aspects of every foodservice operator's job.

Evaluation-related actions taken are important because laws related to why and how employees may be disciplined and/or terminated can be complex. Violation of these laws may cause operators to spend excessive amounts of time defending their actions. These challenges can also create substantial financial hardship if fines or penalties are levied. Also, if widely publicized, violations can result in significant adverse publicity that can negatively affect the operation's sales and profits and the overall morale of its staff members.

The Importance of Regular Performance Appraisal

Effective foodservice operators provide ongoing performance feedback to their work teams, and this process is integral to maximizing the effectiveness of the teams. Those who directly supervise a team member have first-hand knowledge of the member's performance. They are the best choice to perform an employee evaluation, and they are often the employees best able to help others improve their performance levels.

Performance management and employee **appraisals**, when properly planned and implemented, can be useful resources to help team members do their best work.

Performance management is one of an operator's best tools for attracting, retaining, and improving the work of talented team members. Even the smallest foodservice operation can benefit from a formal and ongoing performance management program.

The performance management process involves five key management activities as shown in Figure 11.1.

Key Term

Performance management: A systematic process by which foodservice operators can help employees improve their ability to achieve goals.

Key Term

Appraisal (employee): An objective and comprehensive evaluation of employee effectiveness.

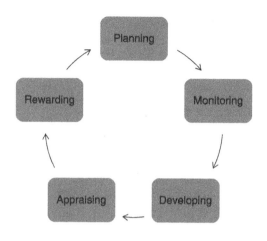

Figure 11.1 The Performance Management Process

✓ *Planning:* Foodservice operators plan the work to be done and set expectations for employee productivity levels. The input of current employees may be helpful in this process.

✓ *Monitoring:* The monitoring of employee performance is a continuous process, and it must be undertaken in a fair and consistent manner.

✓ *Developing:* The development of a team member's skills is also usually a continuous process as existing employees gain greater skills and new employees initially learn skills required to perform their jobs.

✓ *Appraising:* Employee appraisal maybe informal and occur continuously, or it may be formal and conducted on a set schedule. An employee appraisal process must be perceived as fair by all of those being evaluated.

✓ *Rewarding:* All team members expect and appreciate rewards for good performance. The rewards may be financial or non-financial, but they must always be tangible and perceived as valuable by the team.

The distinguishing characteristic of an effective performance management program is its focus on achieving results. It addresses the effectiveness of employees and their work processes and output quality. To illustrate, consider a skilled foodservice employee who chops cabbage quickly and produces a high-quality product. Traditional performance appraisal systems might rate the employee with high scores because of the hard work and efficiency. A performance management system, however, might involve the worker in an objective assessment of effectiveness. Perhaps the cabbage chopping process should be mechanized (new equipment should be purchased), or prechopped cabbage should be purchased. Then the employee's knowledge and skills might be better utilized doing other important tasks. A properly designed performance management system emphasizes goal attainment over employee effort (output).

Traditional performance appraisal systems frequently emphasize employees' negative characteristics. By contrast, performance management systems identify and correct employee weaknesses, they should also be designed to recognize, reinforce, and reward employees' strengths.

The Characteristics of Effective Performance Appraisal

The most successful performance appraisal programs encompass four critical characteristics.

1) *Performance goals are set by operators and team members*—Goals can be short-term or long-term and address numerous issues. They should be specific and measurable where possible (e.g., completion of a specific task within a defined period of time and at an established quality level). Employees may require additional training or other support to meet the performance goals. As workplace changes occur, goals should be reviewed and modified, when needed, with input from employees.
2) *Regular and informal feedback from supervisors is provided*—Traditional annual formal appraisals alone do not normally allow employees to assess progress toward goal attainment. Generally, more frequent input is needed, and this occurs as supervisors work closely with employees and provide them with on-going coaching (see Chapter 9).
3) *A formal method to address performance or disciplinary problems is in place*—Methods used to correct inadequate job performance should be known, be fair, and be applied equally to all employees. From a legal perspective, this means using a formal method that details, in writing, the procedures and policies to be followed by management. Performance problems should be identified as they occur, and a course of action for improvement should be agreed on. Written procedures should require managers to document problems and agree upon problem resolution plans.
4) *Regular and formal appraisals are undertaken*—Formal reviews that accurately document each team member's performance should be conducted regularly. In addition to pinpointing improvement concerns, appraisals should identify specific steps for employee approval including improvement to enhance their long-term position with the operation.

Each of these performance appraisal characteristics may seem obvious. However, busy foodservice operators may think they lack the time or skills required to implement improvements. Despite that, experienced foodservice operators know that the implementation of an effective performance appraisal system will, in the long run, save time and money. It will also improve the quality of work teams and the guest service they provide.

Benefits of Performance Appraisal Programs

A properly implemented formal performance appraisal system yields many benefits, and they include:

✓ *The recognition of outstanding performance*—In the best appraisal systems, employees learn about the areas in which they excel. This increases their morale and helps reduce turnover. Every team member likely has positive personal work characteristics such as attendance, punctuality, neatness, adherence to dress code, friendliness, and/or other traits. Foodservice operators can help build each team member's personal esteem. It is also important to recognize that positive reinforcement of a team member's existing strengths often makes it easier to achieve improvements in other areas.

✓ *The identification of necessary improvements*—When an employee can excel in a position, the job becomes more enjoyable. Few employees want a job that they do not understand, nor one in which they perform poorly. Some employees may not know about necessary improvements. When a fair and unbiased foodservice operator conducts regularly scheduled appraisals, employees will learn how their performance can improve. The result is positive for both the employee and the operation.

✓ *The clarification of work standards*—Well-designed performance appraisal systems emphasize how well employees have attained established goals. Sometimes this is simple. For example, the average speed at which a grill cook produces guest orders can be calculated using ticket start and stop time data from an operation's point-of-sale (POS) system.

In other cases, proper performance can be more difficult to measure. For example, foodservice operators agree that friendliness is an important characteristic of a good host or hostess. However, an objective evaluation of this trait is more complex than a timed measurement of task completion.

Regardless of measurement difficulty, however, if friendliness is evaluated, an employee should understand its importance, how the trait might be displayed within a job, and the expected end results of displaying it. Guest comments, observation by management, and mystery shopper services (see Chapter 9) can provide input for this appraisal. If a friendliness concept is not communicated clearly, whether a foodservice operator can fairly evaluate an employee's efforts to display it is questionable. Effective performance appraisal includes the responsibility of operators to clearly define and communicate job expectations.

✓ *The opportunity to analyze and redesign jobs*—An effective performance appraisal program can identify the need for job redesign. For example, consider a better way to perform taking inventory. Changes in methodology are suggested by an employee who is responding to concerns that their performance of this task is below standards. In this case, the employee's ideas on how to do the job better may result in real benefits to the worker and the operation.

✓ *The identification of specific training and staff development needs*—A performance appraisal system that identifies deficiencies but does not address them is a poor one because it creates frustration. Specific steps that an employee and the operation should take to improve the employee's skill levels are needed. Opportunities to discuss professional development activities can be an important part of this dialogue.

✓ *The determination of professional development activities*—Information discussed during a performance appraisal session may establish a foundation to help plan an employee's career. If career goals are known, beneficial educational and/or training activities can be considered, agreed on, and used as a benchmark for subsequent performance appraisal. For example, an agreement might be made that an employee will complete an online foodservice-related course offered by a community college, and the foodservice operation will reimburse the employee if the course is successfully completed. This could be a factor in a subsequent appraisal session because the operator and the employee have agreed that successful completion of the course would be a priority.

✓ *The validation of the recruiting, screening, and selection processes*—In Chapters 5 and 6, the importance of recruiting, interviewing, and selecting team members was addressed. Formal performance appraisal program sessions allow foodservice operators to evaluate the effectiveness of these procedures. If employees consistently do not meet expectations, the screening and selection procedures in use may be the cause. Well-managed appraisal systems help operators to pinpoint causes of employee selection shortcomings.

✓ *The providing of opportunities for employee feedback and suggestions*—The best foodservice operators use appraisal sessions to learn about issues that affect guest satisfaction from the employees who interact with guests. Foodservice employees in all job classifications serve either as **internal customers** or **external customers**. *In most cases, these workers will be* eager to share their thoughts and beliefs about how their jobs could be improved and how service to their customers could be enhanced.

Key Term

Internal customers: Employees of a foodservice operation.

Key Term

External customers: Guests of a foodservice operation.

✓ *The use of an objective method to identify candidates for pay increases and promotion*—Performance appraisal systems commonly yield decisions about which employees receive pay increases and promotions. It is a workplace reality that scarce organizational resources must be allocated carefully, and properly designed appraisal systems help with this key task.

Popular Performance Appraisal Methods

Those foodservice operators who have determined that a performance appraisal program should be instituted in their businesses can choose from several popular performance appraisal methods. These are:

Absolute Standards
Relative Standards
Targeted Outcomes

Absolute Standards

When an **absolute standard** appraisal method is used, employee performance is compared with an established productivity standard not related to any other employee.

Examples of absolute standard evaluation methods include:

Key Term

Absolute standard (performance appraisal method): Measuring an employee's performance against a previously established standard.

Critical incident. Critical incidents are those specific behaviors needed to do a job successfully. With this approach, behavioral traits that employees exhibit on the job (the critical incidents) are documented in writing. For example, an observation about Tammy, a bartender, might be: "Tammy showed poise, maturity, and patience with an agitated guest after she could no longer serve him alcoholic beverages. She calmed him down, and he ordered coffee."

Note that, in this example, the incident report focuses on Tammy's behaviors and their results, rather than on Tammy's personal traits. One advantage of this method is that, during a formal appraisal session, the discussion can address specific positive and negative critical incidents that have occurred to help support an objective evaluation of the employee's performance. A disadvantage is that frequent documentation of critical incidents can be time consuming.

Checklist appraisal. As illustrated in Figure 11.2, a checklist appraisal system with a "yes" or "no" responses is used to address behavioral factors related to the successful completion of tasks identified in a job description. Note that, in this example, specific tasks unique to a drive-through order taker's job are addressed. While the checklist can be modified to apply to specific positions, this can become a disadvantage if individualized checklists must be prepared for many job categories and positions.

	YES	NO
1) Supervisor's instructions consistently followed	____	____
2) Quality of verbal communication with guests is consistently acceptable	____	____
3) Guest order accuracy is consistently good	____	____
4) Speed of order taking is consistently good	____	____
5) Guest check totals are consistently accurate	____	____
6) Quality of guest problem resolution is consistently good	____	____

Figure 11.2   Drive-through Order Taker Checklist Evaluation

Continuum appraisal. In this approach, a scale is used to measure employee performance relative to specific factors. The point on the scale that best represents the employee's performance is selected. Figure 11.3 illustrates a sample of continuum appraisal questions that address two factors: work quantity and dependability. Note that each performance factor is carefully defined to maximize consistency among those conducting the appraisals.

The number of alternative rating choices in a continuum appraisal system usually ranges between four and eight. In Figure 11.3, each of the five rating response categories, like all continuum scales, represents a ranked level of

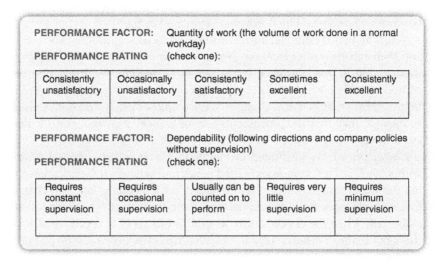

Figure 11.3   Sample Continuum Appraisal Questions

measurement. While the categories represent a sequence (for example, more to less, stronger to weaker, or very unsatisfactory to highly satisfactory), the categories do not indicate the magnitude (extent of differences) between each level. While the system lacks the depth of analysis found in, for example, a critical incident appraisal system, an advantage of continuum appraisal is its ease of use.

Relative Standards

When foodservice operators use a **relative standard** of performance appraisal, they compare one employee's actions with those of another peer doing the same job.

The two most common approaches to this type of appraisal are:

1) Group order ranking
2) Individual ranking

Key Term

Relative standard (performance appraisal method): Measuring one employee's performance against another employee's performance.

1) *Group order ranking.* Group order ranking requires the evaluator to place the employee into a specific classification such as the top 10% or lowest 50%. This approach is often used when evaluating employees for possible promotion or pay increase.

To illustrate, assume that Ramon is the dining room supervisor at a restaurant, and he supervises 20 servers. He must rank (compare with each other) all 20 employees. For example, if the system asks Ramon to identify his top 10% of employees, he must identify only his top two employees (10% of 20 employees = 2 employees).

One advantage of this appraisal system is that raters cannot inflate evaluations, resulting in everyone being rated above average, nor can they rate nearly all employees as average or below average. A disadvantage, especially with small groups of employees, is that some individual or individuals must always be rated in the below-average group, regardless of their actual talent level. A second disadvantage is that a supervisor with clearly inferior subordinates will still generate groups of *best* and *worst* staff members (although the best employees may simply be the best of the worst!). Conversely, a supervisor with a group of outstanding employees must still rank some of them in a below-average group.

2) *Individual ranking.* This method requires supervisors to rank employees in order from highest to lowest. Only one employee can be rated as the very best, and ties (equal ratings) are typically unacceptable. This system tends to work best with smaller groups of workers. However, because this system also represents a "succession" level of measurement, the rank achieved does not indicate

the magnitude (extent) of differences between the ranks. For example, if 10 employees are ranked, there is no real rationale for believing that the difference between the first- and second-ranked employees is equal to the difference between the ninth- and tenth-ranked workers. The system is simple, but it is also subjective, not objective.

Targeted Outcomes

Another approach to employee evaluation involves the identification of **targeted (achieved) outcomes**. In this system, employees are evaluated based on how well they accomplish a specific set of objectives critical to successful job completion. For example, an operation's chef managers might be evaluated primarily on whether they did (or did not) achieve pre-established food cost percentage targets.

Goal setting and goal achievement measurements and rewards are not new concepts. For example, **management by objectives (MBO)**, the concept of using identifiable objectives to measure performance and to assign employee rewards, is decades old.

Essentially, a targeted outcome appraisal system requires four components:

1) Identification of potential performance targets
2) Employee input in final target selection
3) A defined period of time for target completion or achievement
4) Performance feedback

Properly managed, targeted outcome appraisal systems can be successfully implemented at all levels of a foodservice operation. Regardless of the performance appraisal system put in place, however, foodservice operators must understand some challenges faced when measuring employee performance. Figure 11.4 presents seven significant threats (including **halo effect** and **pitchfork effect**) to legitimate employee appraisal and specific examples of their occurrence.

Key Term

Targeted outcomes (performance appraisal method): Measuring the extent to which specified performance goals were achieved.

Key Term

Management by objectives (MBO): A plan developed by an employee and their supervisor that defines specific goals, tactics to achieve them, and corrective actions, if needed.

Key Term

Halo effect: The tendency to let the positive assessment of one individual trait influence the evaluation of other traits that are unrelated.

Key Term

Pitchfork effect: The tendency to allow the negative assessment of one individual trait to influence the evaluation of other, nonrelated traits.

1) Basing valuation scores on an employee's most recent behavior rather than evaluating the entire performance. For example, rating a usually outstanding employee with negative remarks based on a recent argument.

2) Allowing irrelevant or non-job-related factors to influence the evaluation. For example, evaluating physical appearance or disabilities, race, social standing, participation in employee assistance programs, or use of excused time off instead of actual performance.

3) Failing to include unfavorable comments on the evaluation, even if justified. For example, not wishing to offend an employee by not discussing undesirable traits such as poor personal grooming or the consistent display of a negative attitude.

4) Rating all team members at about the same point on a ranking scale (usually the middle). For example, the tendency of supervisors, because they want to be liked by all employees, to avoid strong negative (or even positive) statements about their workers.

5) Judging all employees too leniently or too harshly. For example, the tendency of some supervisors wanting to enhance their own credibility to unfairly criticize (or praise) the performance of those they are evaluating.

6) Permitting personal feelings to bias the evaluation process. For example, the tendency of some supervisors to rate employees they like very highly, while rating those employees they dislike much lower.

7) Allowing one very good (or very bad) trait to affect all other ratings of the employee (e.g., the halo effect or the pitchfork effect). For example, rating an employee with one exceptional (or unfavorable) trait as equally exceptional or unfavorable in all other measured traits.

Figure 11.4 Threats to Fair Performance Appraisals

Technology at Work

The primary purpose of any performance evaluation is to measure job performance and recognize and reward top employees for a job well done.

Employee performance management software products available in the market today generally offer features for setting employee goals and expectations, tracking employee performance, coaching, and conducting one-on-one performance appraisal meetings. Some products can also collect feedback and recognize employees' accomplishments with rewards. After gathering performance management data, foodservice operators can make more informed decisions about key issues such as employee compensation and career advancement.

To examine some of the products that can assist foodservice operators in managing their employee performance appraisal programs, enter "employee performance appraisal software for restaurants" in your favorite search engine and review the results.

What Would You Do? 11.1

"This just doesn't make any sense," thought Jim Cramer, the director of operations for the Bartello Palms Retirement Center.

Jim was reading an employee performance appraisal report written by Ron Dukes. The employee Mr. Dukes evaluated was Sandy Rinaldo. Sandy had worked in the retirement center's kitchen for over 10 years, but Ron had only recently been appointed as her supervisor, and this was his first formal evaluation of Sandy.

In prior evaluations, Sandy had consistently received outstanding reviews. However, the report Jim was now reading indicated that Sandy's current work performance was far below average and that her attitude was poor. In fact, Ron was recommending immediate disciplinary action for Sandy's failure to meet his performance expectations.

Jim knew that the best employee performance appraisal systems provided ratings based on employee performance and not on the personal opinions of the supervisor assessing the employee. Perhaps Sandy's performance had declined. Jim knew it was also possible, however, that the problem was not with Sandy, but with Ron Dukes—and with the ability of the performance appraisal system to be consistently and reliably applied regardless of who was doing the appraisal.

Assume you were Jim Cramer. How would you go about assessing whether this employee's evaluation was performed fairly? How important would it be that you do so?

Progressive Discipline Programs

Discipline is a foodservice operator's response to an employee's improper behavior. In this sense, the term discipline is applied as in the disciplining of a child (e.g., as a punishment for their behavior or to prevent future improper behavior).

A better use of the term "discipline," however, is utilized by the military. A **disciplined** group in the military is a squad or platoon in which soldiers follow orders and perform in a way that enhances the ability of the unit to achieve its objectives. In a similar manner, a disciplined work team in the foodservice industry is one in

Key Term

Discipline: Any effort designed to influence an employee's negative behavior.

Key Term

Disciplined (work team): The situation in which employees conduct themselves according to accepted rules and standards of conduct.

which team members follow the organization's established set of rules, polices, and procedures. As they do this, they best serve guests and meet the operation's financial objectives.

In most cases, foodservice operators find that using a formal **progressive discipline program** will be of great value both to them and to their employees.

A proper progressive discipline program as utilized in the foodservice industry typically consists of four steps (phases) as shown in Figure 11.5.

In a progressive disciplinary program, **positive discipline** is used to encourage desired behavior, and **negative discipline** is used to discourage improper behavior.

Foodservice operators typically use direct instruction, written directions, employee manuals, role modeling, and company traditions to relay their expectations about employee behavior. Most employees will adhere to specific behavioral standards when they know what is expected of them.

Key Term

Progressive discipline program: program designed to modify employee behavior through a series of increasingly severe punishments for unacceptable behavior.

Key Term

Positive discipline: Any action designed to encourage desired employee behavior.

Key Term

Negative discipline: Any action designed to correct undesirable employee behavior.

Unintended mistakes and occasional errors do occur, however, and coaching activities can usually correct these actions. However, intentional and repetitive noncompliance with standards should result in pre-established consequences.

A consistently and fairly applied progressive discipline program lets employees know, in advance, the consequences of unacceptable behavior, and they become more serious as the behavior is repeated. The implementation of a progressive discipline program is intended to communicate the consequences of inappropriate behavior.

For this approach to be effective, the worker must view the outcomes of repeated behavior to be undesirable. If they are not, the behavior is unlikely to change.

Figure 11.5 Steps in an Effective Progressive Disciplinary Program

Documented Oral Warning

The first (mildest) step in a progressive disciplinary program is the **documented oral warning**. The written record of an oral warning should include the employee action that caused the warning, the date of the incident and of the oral warning, and the name of the supervisor issuing the reprimand.

Key Term

Documented oral warning: The first step in a progressive discipline process: a written record of an oral reprimand given to an employee is provided.

Not every violation of a policy or procedure should result in a documented oral warning. For example, simple coaching to correct behavior may come before the documented oral warning. This action may prove successful, especially if a one-time occurrence does not create significant difficulties (for example, an initial policy violation about the necessity of wearing solid-toed shoes in the kitchen).

Some foodservice operators believe that an initial discipline activity should not become part of the employee's permanent personal file. Instead, a form can be maintained in a separate manager's file. Other managers document the issuance of an oral warning, but they do not include the actual reprimand in the employee's permanent file. The property's progressive discipline program should explain how oral warnings must be documented so all employees receive identical treatment.

To illustrate this initial progressive discipline step, assume that Saul, a new cook in a hospital's kitchen has been late twice in the past week and has been coached about the problem by Ajay, his supervisor. When the problem next occurs, Ajay meets privately with Saul to explain why punctuality is important. Ajay should then allow Saul to respond. Ajay may discover that Saul did not understand some aspect of the expected behavior, or he (Ajay) might learn about a legitimate reason for the late arrival.

Ajay and Saul should mutually develop and agree on an appropriate solution, and Ajay should tell Saul about the consequences of further late arrivals. An oral reprimand record can then be completed and signed by Ajay and Saul to finalize the oral warning process. Hopefully, this will resolve the problem. If not, the next step in the progressive disciplinary process will be necessary.

Written Warning

A **written warning** is a document that becomes part of an employee's permanent file. Its content is like that of an oral written warning. The written warning should include the employee's name, date of incident, name of the supervisor

Key Term

Written warning: The second step in a progressive discipline process that alerts an employee that further inappropriate behavior will lead to suspension.

issuing the written warning, and the plan to prevent further occurrences. Written warnings also typically permit the employee to provide their own version of the incident. The signatures of the supervisor and the affected employee are required.

The written warning step must be correctly applied to protect the organization if the employee later challenges the legality of the progressive disciplinary process. The employee behavior leading to this step should relate to that exhibited in the first step. Managers should remember that the courts generally view the term, *progressive discipline,* as discipline "related to the same issue."

Returning to the example of Ajay and Saul, assume that Saul neglected to properly label and date a pan of food before placing it in the cooler, and this action resulted in the loss of the product. It would not be appropriate to issue a written warning to Sam for this behavior, because it does not relate to his late arrivals. Since this is a first offense, a documented oral warning would be appropriate. Sometimes it is not so clear whether an incident is one in a series or the beginning of a new series of incidents. When that is the case, this question must be carefully answered to determine the appropriate management action.

While it is generally a good rule to praise in public and to reprimand in private, many progressive discipline processes require an observer to be present at the second and later steps in the process. Owners, co-managers, supervisors, or others can monitor the discussion and serve as an eyewitness.

Suspension

Suspension is the third step in the progressive discipline process, and it should be undertaken if the previous two steps have not resolved the performance problem. A suspension may be for any period deemed appropriate by management. Suspensions, however, should be applied consistently: if one employee is suspended for a specific period for a particular behavior, then all other employees suspended for the same negative behavior should be suspended for the same length of time.

Key Term

Suspension: The third step in a progressive discipline process; a (typically unpaid) period off from work resulting from on-going inappropriate behavior.

An action to suspend an employee must be documented, and the information should be placed in the employee's permanent file. Some employees may refuse to sign the suspension document. Despite an employee's view that an unsigned document will somehow invalidate it, this refusal carries virtually no legal meaning if the employee was given an earlier opportunity to sign the document.

If an employee refuses to sign a discipline report, the observer should sign the document and note the employee's refusal to sign. While most suspensions result in un-paid time off, an employee's suspension, with or without pay, is a serious step. It should signal that the employee's behavior is clearly not acceptable, and that they are in danger of losing their position with the organization. If the employee's behavior is not corrected after this step, dismissal is likely.

Dismissal

Dismissal is the final step in the progressive discipline process and should be implemented only for continued serious infractions. Unfortunately, in some cases this step may be a foodservice operator's only alternative. Later in this chapter, you'll learn more about the specific issues related to

Key Term

Dismissal: An employer-initiated separation of employment.

employer-initiated employee separations. In the context of a progressive disciplinary program, however, dismissal represents not only a final step in the process but also a failure on the part of both the employee and the employer. This is so because an effective progressive discipline program should modify and improve employee behavior.

From an employee's perspective, dismissal means that, even after repeated warnings, the organization's behavioral expectations were not met. From the employer's perspective, dismissal means that the manager was unable to persuade the employee to modify their behavior sufficiently to maintain their job.

Find Out More

The use of a progressive discipline program is always a good idea. However, in some cases an employee termination must take place immediately and bypass the previous steps in the program. On-the-spot firing cannot only occur, in some cases it *must* occur to protect an employer from potential liability.

Actions or violations of specific polices—if sufficiently egregious—may create substantial legal exposure for a foodservice operator. The reason: failure to act may pose a risk of potentially serious physical and emotional harm to employers, employees, or others. Then a failure to terminate an employee could result in charges of negligent retention (see Chapter 6).

Some examples of misconduct that might justify or require immediate termination include possession of weapons in violation of company policy, falsification or destruction of company records, use of illegal drugs while working, selling or distributing illegal substances while working, and a manager's retaliation against an employee who made an internal harassment complaint.

To gain a better understanding of behavior that should result in the immediate separation of a worker's employment, enter "legal grounds for immediate employee termination" in your favorite search engine and view the results.

Technology at Work

Employee discipline and correction may be needed when team members do not follow a foodservice operation's rules and/or when they display behavioral problems that affect their job performance or impact others negatively.

Discipline issues can range from being late consistently to harassing coworkers and more. Just as special software can help foodservice operators analyze their sales, special software programs can give foodservice operators the ability to track employee discipline problems and resolve them efficiently.

Good employee discipline software can increase productivity by automating tasks like documenting oral warnings, recording suspensions, and initiating legally allowable terminations.

To learn more about special software designed specifically for the administration of progressive discipline programs, enter "progressive discipline program software" in your favorite search engine and review the results.

Staff Separation

In some cases, a progressive disciplinary program results in the separation of a team member. Despite the known costs and disruption that occur when staff members leave their jobs, some foodservice operations continually experience very high employee turnover rates (see Chapter 1). The financial cost of employee turnover is real. In one study, the Center for Hospitality Research at Cornell University estimated that the actual cost of employee turnover averages around $5,864 for the typical front-line staff member,[1] as detailed in Figure 11.6.

These above costs likely increase every year. However, regardless of the exact amounts, experienced operators know costs associated with replacing employees who do not remain with the organization are high. They should be avoided, when possible, so operators will not need to continually recruit new staff members and

Cost Category	Actual Cost $
Pre-departure costs of previous employee	$ 176
Recruiting new employee	1,173
Selection of new employee	645
Orientation and training of new employee	821
Productivity loss until new staff member is experienced	3,049
Total	**$5,864**

Figure 11.6 Average Cost of Front-line Restaurant Employee Turnover

---

1 https://www.notch.financial/blog/restaurant-turnover-rate. Retrieved June 30, 2023.

move them through the expensive and time-consuming orientation and training processes.

It is also important to recognize that, in geographical areas with tight labor markets, significant loss of staff can sometimes mean reducing the number of operating hours or choices of menu items to be served. When these issues occur, actual costs are incurred, and revenue is also reduced.

Of course, some turnover is inevitable and can even be good for a business because, in many cases, new staff members with diverse skills, attitudes, and ideas can be recruited to fill vacant positions. However, excessive employee separations and high turnover rates that result are detrimental to an operation's ability to maintain quality standards and costs and, sometimes, to remain financially viable. Employee separations can be viewed as **voluntary** or **involuntary**.

Key Term

Voluntary (separation): An employee-initiated termination of employment.

Key Term

Involuntary (separation): An employer-initiated termination of employment.

Voluntary Separation

Voluntary employment separations are often inevitable for several reasons. For example, foodservice operations hire the highest percentage of teenagers of any service industry. This means that, for many front-line workers, a job in the foodservice industry is their first, and some will inevitably look for employment in a different industry. As well, in many communities there will be several foodservice operations to select from when deciding where to work. If a talented staff member hears that workers doing the same job are making more money in another operation it is often easy for them to be hired in that operation. Even highly satisfied team members graduate from school, retire, move away, or, for numerous other reasons, resign from a job.

While these employee-initiated separations are often inconvenient, they rarely cause significant replacement issues. In the best-case scenario, employees will inform managers about their pending departures with sufficient time for replacements to be recruited and trained.

Involuntary Separation

As previously addressed, involuntary employment separations are frequently caused by poor employee performance. They are documented in a progressive discipline program and result in a required dismissal. However, foodservice operators may also have failed to properly select, orient, train, and/or direct the work of these employees.

In most cases, employers can justify the involuntary separation of their employees based on the at-will employment doctrine (see Chapter 6). At-will employment laws usually allow employers great latitude to terminate employees with or without cause. However, there are at least five exceptions to these laws. If evidence that any one of these exceptions is documented in the employee appraisal or progressive disciplinary program, the affected employee could likely win a wrongful termination lawsuit. These exceptions are:

1) *Contractual relationship*—Employees may not be terminated at will if there is a contractual relationship. A contractual relationship exists when employers and employees have a legal agreement regarding how employee issues are handled. Under these contracts, discharge may occur only if it is based on **just cause**.

> **Key Term**
>
> **Just cause:** A legally sufficient reason. Also sometimes referred to as good cause, lawful cause, or sufficient cause.

2) *Implied contractual relationship*—An implied contract is any verbal or written statement made by members of the organization that suggests organizational guarantees or promises about continued employment. These statements are most often found in employee handbooks that have not been carefully reviewed to ensure there is no implied contractual relationship.

3) *Public policy violation*—An employee cannot be terminated for refusing to obey an order from an employer that is considered illegal activity (e.g., being asked to bribe a food safety inspector to receive a higher kitchen inspection score). Furthermore, employees cannot be terminated for exercising their individual legal rights. Examples can include agreeing to serve on jury duty and filing a complaint against an employer with a governmental entity (even if the complaint is dismissed because it was unfounded).

4) *Statutory considerations*—Employees may not be terminated if doing so would result in a direct violation of a federal or state statute. An employer may not terminate an employee for reasons that would violate either the Civil Rights or Age Discrimination Acts. For example, an employer who terminates a female employee in preference to a male employee is in violation of a federal mandate that such gender-based employment decisions are not permitted.

> **Key Term**
>
> **Good faith:** A concept that requires parties to an agreement to exercise their powers reasonably and not arbitrarily or for some irrelevant purpose. Certain conduct may lack good faith if one party acts dishonestly or fails to have concern for the legitimate interests of the other party.

5) *Breach of good faith*—It is difficult for employees to prove breach of **good faith** on the part of an employer. However, courts have determined

that the at-will employment concept does not allow employers total freedom to terminate. For example, assume a commission-based catering salesperson secures a contract for several catered events. The employer cannot terminate that person's employment to avoid paying the large commissions rightly due to the salesperson. In this case, the courts would assume the employee worked on the sale in good faith and expected to earn the commission. The employer cannot use the at-will employment doctrine to avoid paying commissions because doing so would not be acting in good faith.

Involuntary employee separations affect more than just immediate employee replacement and training costs. In most states, employees who are involuntarily separated qualify for unemployment compensation payments. While there are exceptions (e.g., for employees dismissed for theft, assault, or other illegal activities), significant increases in payments to employees who are involuntarily separated result in an increase in the amount the employer must pay into their state's unemployment compensation accounts.

Unemployment insurance programs are designed to provide laid off or terminated workers with weekly income during short periods of unemployment. They are funded by both State Unemployment Tax Act (SUTA) funds and the Federal Unemployment Tax Act (FUTA) funds paid into by employers.

In most states, unemployment taxes are calculated based on a foodservice operator's size of operation, the dollar amount the operation has paid in wages, and the unemployment insurance benefits collected by former employees. These factors all contribute to what is commonly referred to as the **employers' experience rating**.

Key Term

Employers' experience rating: An assessment of unemployment insurance taxes paid by employers who have paid covered wages for a sufficient period to rate their experience with unemployment insurance. In most cases, the less unemployment an employer's workers have experienced, the lower the employer's unemployment insurance tax rate will be.

Find Out More

After making a layoff or termination, many foodservice operators ask the following question: "What will happen to my Unemployment Insurance (UI) tax rate?" and rarely can a sufficient answer be provided.

The tax impact of a layoff or qualifying termination depends on several additional factors, all of which interact with one another at the same time. Among the factors are the number of layoffs previously made by the employer, the period for which each laid-off employee will collect UI benefits, the amount of money in the UI trust fund of the employer's state, and even the number of layoffs and terminations made by other employers!

Despite the complexity of the issue, it is a good idea for a foodservice operator to understand as fully as possible the impact of employee separation on their own UI tax rates. To find out more about how employee separations could impact a foodservice operator's state unemployment tax in a specific state, enter "factors that affect SUTA rates in (name of state)" in your favorite search engine and review the results.

Exit Interviews

Separation of employment may be voluntary or involuntary, but there is a third category of separation that operators must also understand. This is the employee who leaves *voluntarily* but does so *involuntarily*! In other words, the employee decides to leave the job, but the reason relates directly to a negative aspect of the job.

To illustrate, consider Cheri, a good employee who likes her foodservice job but who is leaving for another job doing the same work at the same rate of pay. She currently feels her manager does not appreciate her, and recognition of a job well-done is important to her (as it is to many employees). Since Cheri does not receive the recognition she seeks, she is leaving the organization. Often, the motivations of employees like Cheri remain unknown. In other cases, exit interviews (see Chapter 8) can help uncover reasons for the involuntary/voluntary separation of employees.

Exit interviews are typically used when an employee voluntarily resigns. Then a foodservice operator may ask questions while taking notes or request that the employee complete a questionnaire or a short survey. Exit interviews can often yield vital information about the workplace and sometimes can prevent the loss of employees who really want to remain at the job but feel they cannot. Figure 11.7 lists examples of questions that may be asked of separating employees during a well-planned exit interview.

The proper acquisition, assessment, and development of a foodservice team is a critical management activity. The quality with which these tasks are completed will directly affect guest service levels and product quality, but it also has a direct financial impact. The management and control of costs related to labor acquisition, development, and retention is so important this process will be the sole topic of the next chapter.

1) What is your primary reason for leaving?
2) Did any specific event or thing make you decide to leave?
3) What did you like best about working here?
4) What did you like least about working here?
5) Did you receive sufficient training for your job?
6) Did you receive adequate feedback about your performance from your supervisor?
7) Did this job help you advance in your long-term career goals?
8) Are there any current employees you feel would perform well in your job?
9) Do you have any tips to help us find your replacement?
10) Do you believe the pay, benefits, and other incentives in our operation were fairly administered?
11) Was the quality of your supervision adequate?
12) Did any company policies or procedures make your job unusually difficult?
13) Would you be leaving this job if your pay were higher?
14) Would you recommend working for this operation to your family and friends?
15) Would you ever consider working with us again in the future?

Figure 11.7   Sample Exit Interview Questions

What Would You Do? 11.2

"Let me see if I understand," said Joseph Meier, the Food and Beverage Director at Foxwoods Country Club, as he reviewed the employee file of Mariana Aguilar, a dining room server. "Mariana has worked here for four years with no write-ups. Now, in the two weeks since you have become her supervisor, Mariana is at stage three of our four-stage progressive discipline process, and your recommendation is suspension."

"She did it to herself, Joe," replied Steven, a former dining room server and Mariana's new boss. "Mariana never liked me. I think she is jealous of me because I got promoted, but I played it straight up. Every write-up I give is legitimate. She needs to be suspended or fired because I don't think she'll ever change."

Joseph reviewed Mariana's file one more time. Mariana' first written oral warning, dated 10 days ago, was for returning from her unpaid lunch period 10 minutes late. Mariana stated that she had gone to the bank and was caught at a train crossing returning to work, which delayed her timely return.

The written warning followed two days later, when Mariana was reprimanded for being out of uniform. In fact, in the written report, Mariana admitted that she was working and had forgotten to put on her nametag when she clocked in. However, she had been clocked in for only five minutes and was on her way back to the locker room to put on her name tag, when a club member stopped her and asked her some questions. While Mariana was responding to the member, Steven walked by and noticed Mariana was "working while out of uniform." Steven then verified that Mariana had already punched in, and, as a result, he wrote her up.

The suspension that Steven was now proposing was the result of an incident yesterday when, as required by the employee handbook, Mariana failed to notify her supervisor four hours before her shift began that she would not be at work as assigned on the schedule. On that day Mariana was scheduled to work at 7:00 AM, but at 3:00 AM had to take her 2-year-old son to the emergency room for a condition diagnosed as a severe ear infection.

At 5:00 AM Marianna had left a voice mail on Steven's cell phone stating she was still at the hospital and would not be in that day. That 5:00 AM call was placed two hours later, Steven now pointed out, than the four-hour notice required by the employee manual, and so it justified the suspension.

Assume you were Joseph: Should you support Steven's recommendation that Mariana be suspended? Do you think Steven is properly following your organization's progressive discipline program? Explain your answer.

Key Terms

Performance
  management
Appraisal
  (employee)
Internal customers
External customers
Absolute standard
  (performance
  appraisal method)
Relative standard
  (performance
  appraisal method)

Targeted outcomes
  (performance
  appraisal method)
Management by
  objectives (MBO)
Halo effect
Pitchfork effect
Discipline
Disciplined (work team)
Progressive discipline
  program
Positive discipline

Negative discipline
Documented
  oral warning
Written warning
Suspension
Dismissal
Voluntary (separation)
Involuntary (separation)
Just cause
Good faith
Employers' experience
  rating

Operator's 10-Point Tactics for Success Checklist

Evaluate your need for, and the current status of, each of the following operational tactics. For those tactics you think are important, but not yet in place, develop an action plan for its implementation including who will be responsible for the tactic's completion and the target date by which it should be completed.

Tactic	Don't Agree (Not Done)	Agree (Done)	Agree (Not Done)	If Not Done Who Is Responsible?	Target Completion Date
1) Operator recognizes the importance of regular performance appraisal in the development of a professional foodservice staff.	——	——	——		
2) Operator can state the features of an absolute standard performance appraisal system.	——	——	——		
3) Operator can state the features of a relative standard performance appraisal system.	——	——	——		
4) Operator can state the features of a targeted outcome performance appraisal system.	——	——	——		
5) Operator recognizes the purpose of a "documented oral warning," the first step in a progressive discipline program.	——	——	——		
6) Operator recognizes the purpose of "written warnings," the second step in a progressive discipline program.	——	——	——		
7) Operator recognizes the purpose of "suspension," the third step in a progressive discipline program.	——	——	——		

Tactic	Don't Agree (Not Done)	Agree (Done)	Agree (Not Done)	If Not Done	
				Who Is Responsible?	Target Completion Date
8) Operator recognizes the importance of "dismissal," the fourth and final step in a progressive disciplinary program.	_____	_____	_____		
9) Operator can describe the difference between voluntary and involuntary staff separations.	_____	_____	_____		
10) Operator recognizes the potential value of conducting exit interviews with departing staff members.	_____	_____	_____		

12

Managing Labor-Related Costs

What You Will Learn

1) The Importance of Managing Labor-Related Costs
2) How to Control Total Labor Costs
3) How to Evaluate Labor Productivity

Operator's Brief

In this chapter, you will learn about controlling labor-related costs, an expense that, for most foodservice operations, is second in amount only to that of prod-uct food and beverage) costs. In some operations, labor costs actually exceed the amount spent on food and beverage products.

A foodservice operation's total labor cost is affected by several factors including:

1) Employee selection
2) Training
3) Supervision
4) Scheduling
5) Breaks
6) Menu
7) Equipment/tools
8) Service level provided

Each of these key labor cost-related factors is addressed in detail in this chapter.

Payroll costs that represent the actual cash payments required to maintain your workforce consists of three key components. When utilizing the Uniform

System of Accounts for Restaurants (USAR) to prepare an income statement, these components are listed on the statement as:

1) Management
2) Staff
3) Employee benefits

The sum of these three components is also listed on the income statement and identified as "Total Labor." In this chapter, you will learn how each of these three categories can be properly managed to optimize your total labor cost expenditures.

Labor costs can be classified in several ways including management and staff, fixed and variable payroll, and controllable and non-controllable labor expenses. Each of these classifications and understanding why each is important to effective cost control are discussed in this chapter.

To determine how much labor expense is needed to operate your business, productivity standards must be established. There are a variety of productivity measures currently in use in the foodservice industry including Sales per Labor Hour, Labor Dollars per Guest Served, and the Labor Cost Percentage. Each of these productivity measures are addressed in detail in the chapter including an examination of how they are calculated and their strengths and potential limitations.

Finally, this chapter concludes with an examination of specific steps you can take to reduce your labor-related costs if you find them to be excessive.

CHAPTER OUTLINE

The Importance of Labor Cost Control
    Factors Affecting Total Labor Costs
    Total Labor Costs Reported on the Income Statement
Managing and Controlling Total Labor Costs
    Types of Labor Costs
    Controlling Total Labor Costs
Assessment of Total Labor Costs
    Sales per Labor Hour
    Labor Dollars per Guest Served
    Labor Cost Percentage

The Importance of Labor Cost Control

In years past, labor in a foodservice operation was relatively inexpensive. Today, however, in an increasingly costly labor market, foodservice operators must utilize various techniques to optimize their labor cost and to maximize staff effectiveness. They must also apply cost control tools to evaluate their staff's productivity.

In some sectors of the foodservice industry, a reputation for long hours, poor pay, and undesirable working conditions has caused some high-quality employees to look elsewhere for more satisfactory careers. However, it does not have to be that way. When labor costs are adequately controlled, foodservice operators will have the funds necessary to create desirable working conditions and pay wages to attract the very best employees. In every business, including foodservice, better employees mean better guest service and, ultimately, better sales and higher profit levels.

Factors Affecting Total Labor Costs

Experienced foodservice operators recognize that their total labor-related costs are directly affected by several important factors that include:

1) Employee selection
2) Training
3) Supervision
4) Scheduling
5) Breaks
6) Menu
7) Equipment/tools
8) Service level provided

Employee Selection

Choosing the right employee for a vacant position in a foodservice operation is vitally important in developing a highly productive workforce. The best foodservice operators know that proper employee selection procedures help to establish the kind of workforce that is both efficient and cost-effective. This involves matching the right employee with the right job, and the process begins with the development of job descriptions and specifications (see Chapter 4).

Job descriptions and specifications are important because they enable foodservice operators to hire employees who are qualified to do a job and do it well. Qualified workers can complete their tasks in a more cost-effective manner than will those who are not qualified. In addition, advertising job vacancies can be expensive and making employee selection decisions can be time consuming. When foodservice operators can choose the right employee the first time, they reduce their employee recruitment and selection-related labor costs.

Training

Perhaps no area under an operator's direct control holds greater promise for increased employee productivity and reducing labor costs than effective training.

In too many cases, however, training in the hospitality industry is poor or almost nonexistent. Highly productive employees are usually well-trained employees, and frequently employees with low productivity have been poorly trained. Every position in a foodservice operation should have a specific, well-developed, and ongoing training plan (see Chapter 8).

Effective training improves job satisfaction and gives trainees a sense of well-being and accomplishment. These factors will also reduce confusion and product waste and eliminate poor service and loss of guests. In addition, supervisors find that a well-trained workforce is easier to manage than one in which employees are poorly trained.

An operator's training programs need not be elaborate, but they must be consistent and continual. Foodservice employees can be trained in many areas. Skills training allows production employees to understand an operation's menu items and how they are best prepared. Service-related training may be undertaken, for example, to teach new employees how arriving guests should be greeted and properly seated. Additional examples are how drive-through and/or pick-up orders should be entered in an operation's point-of-sale (POS) system.

It is important to recognize that training must be ongoing to be effective. Employees who are well trained in an operation's policies and procedures should be reminded and updated as needed if their skill and knowledge levels are to remain high. Effective training costs a small amount of time in the short run but can pay off extremely well in labor-related dollar savings in the longer term.

Supervision

All employees require proper supervision, but not all employees want to be told what to do. Proper supervision means assisting employees to improve productivity. In this sense, the supervisor is a coach and facilitator who provides employee assistance. Supervising should be a matter of assisting employees to do their best, not just identifying their shortcomings. It is said that employees think one of two things when they see their boss approaching:

1) Here comes help!

or

2) Here comes trouble!

For supervisors whose employees feel that the boss is an asset to their daily routine, productivity gains can be remarkable. Supervisors who envision their positions as one of exercising power or who see themselves as taskmasters, rarely maintain the quality of workforce needed in today's competitive marketplace.

It is important to remember that in nearly all cases it is employees (not management) that directly serve guests. When supervision is geared toward helping employees, guests and the entire operation benefit. When employees know that management is committed to providing high-quality menu items and guest service, these team members will perform well and will assist their peers in delivering that level of quality. Also, worker productivity will be optimized, and labor costs will be reduced.

Scheduling

Even with highly productive employees, poor employee scheduling (see Chapter 10) by management can result in increased turnover, low productivity rates, and increased costs.

Consider the example in Figure 12.1, where an operator is considering two alternative schedules for pot washers for an operation that is open for three meals each day.

In Schedule A, four employees are scheduled for 32 hours at a rate of $15.00 per hour. The pot washer's cost in this case is $480 per day (32 hours/day × $15.00/ hour = $480 per day). Each shift (breakfast, lunch, and dinner) has two employees scheduled.

In Schedule B, three employees are scheduled for 24 hours. At the same rate of $15.00 per hour, the pot washer's cost is $360 per day (24 hours per day × $15.00 per hour = $360 per day). Staffing costs in this case are reduced by $120 ($480 − $360 = $120), and further savings will be realized because of reduced employment taxes, benefits, employee meal costs, and other labor-related expenses.

Schedule A assumes that the amount of work to be done is identical at all times of the day. Schedule B covers both the lunch and the dinner shifts with two

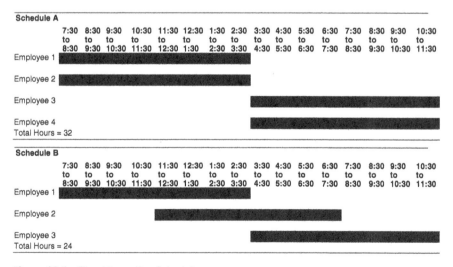

Figure 12.1 Two Alternative Schedules

employees, and it also assumes that one pot washer is sufficient in the early-morning period and later in the day.

Scheduling employee efficiency during the day can often be improved using a **split-shift**, which is a technique used to match individual employee work shifts with peaks and valleys of customer demand. When utilizing split-shifts, an operator would, for example, require an employee to work a busy lunch period, be off in the afternoon, and then return to work for the busy dinner period.

Increasingly, foodservice operators can utilize cloud-based scheduling systems designed to allow them to place the right number of workers in the right shifts on the right days and immediately communicate that information to employees with access to the system. Most of these systems also include features that, with management's pre-approval, allow employees to pick-up, drop, or swap shifts with other workers. These adaptions give staff more flexibility about when and how much they will work. This type of work flexibility is highly valued by employees and can reduce employee turnover caused by inconvenient worker scheduling. As a result, foodservice operators can also reduce their costs related to staff recruiting and new employee training.

Technology at Work

In most foodservice operations, the need for hourly paid employees can vary significantly at different times of the day and on different days of the week. Therefore, developing and maintaining employee schedules is a critical management task.

The scheduling of hourly employees can be time consuming and challenging as operators plan the employees who should work and when they should work. This is especially so when the operation employs many part-time workers with varying scheduling needs.

Fortunately, some companies have designed software programs that assist operators in creating and quickly (and legally!) modifying hourly worker schedules. The best of these programs allows employees to access the schedule from their own smart devices. Then there is no need to return to the operation to learn if a staff member's shift has been added to, dropped from, or modified on the work schedule.

Employee shift scheduling software saves time and ensures employees are always scheduled according to the operation's needs and the employees' personal preferences and availability. Basic employee scheduling programs are

---

(Continued)

often offered at no cost, and those with more advanced features are available for lease.

To review the features of free-to-use employee scheduling tools, enter "best free employee scheduling software for restaurants" in your favorite search engine and review the results.

Breaks

Most employees cannot work at top speed for eight consecutive hours, and they have both a physical and mental need for work breaks. Scheduled short breaks allow them to pause, collect their thoughts, converse with their fellow employees, and prepare for the next work session. Employees given these short breaks will likely produce more than others who are not given any breaks.

Federal law does not mandate that all employees be given breaks, but some states do. Therefore, foodservice operators often must determine both the best frequency and length of designated breaks. Sometimes (and especially regarding the employment of students and minors) both federal and state laws may mandate special workplace break requirements. Professional operators must be familiar with details about these laws if they apply to their businesses.

Menu

A major factor in employee productivity and labor costs is an operation's actual menu. The menu items to be served often have a significant effect on employees' ability to produce the items quickly and efficiently.

In most cases, the greater the number of menu items a kitchen must produce, the less efficient that kitchen's staff will be. Of course, if management does not provide guests with enough choices, loss of sales may result. Clearly, neither too many nor too few menu choices should be offered. The question for operators most often is, "How many selections are too many?" The answer depends on the operation, its employees' skill levels, and the variety of menu items operators believe is necessary to properly attract and serve their guests. *Note*: Franchised operations will, in most cases, be required to offer those menu items dictated by the terms of their franchise agreements.

It is extremely important that the menu items selected by management are those items that can be prepared efficiently and served promptly. If they are, worker productivity rates will be high, and so will guest satisfaction.

Equipment/Tools

Foodservice productivity ratios have not increased as much in recent years as have those of other businesses because foodservice is a labor-intensive rather than a machine or technology-intensive industry.

In some cases, equipment improvements have made kitchen work easier. For example, slicers, choppers, and mixers have replaced human labor with mechanical labor. However, in most cases robotics and automation still do not significantly contribute to the foodservice industry.

Nonetheless, it is critical for operators to understand the importance of a properly equipped workplace and how it improves productivity. This can be as simple as understanding that a sharp knife cuts more safely, quickly, and better than a dull one. It can also be more complex, such as deciding which internet system best provides data and communication links to all stores in a quick-service restaurant chain. In all cases, it is an important part of every foodservice operator's job to provide employees with the tools needed to do their jobs quickly and effectively.

Service Level Provided

The average quick-service restaurant (QSR) employee normally serves more guests in an hour than the fastest server at an exclusive fine dining restaurant. The reason: QSR guests desire speed, not extended levels of employee-provided services. In contrast, fine dining guests typically expect more elegant and personal service that is delivered at a much higher level, and this increases the number of necessary employees.

After operators fully recognize the non-cash payment factors that impact their total labor costs, they must next understand the resulting cash payments that will comprise their reported labor costs. These are reported on an operation's income statement.

Total Labor Costs Reported on the Income Statement

A foodservice operation's actual revenue, expense, and profits for a particular time period are summarized and reported on its **income statement**.

While, in most cases, there are no laws dictating how a foodservice operation reports its financial data, many operators utilize the **Uniform System of Accounts for Restaurants (USAR)** to do so. Now published in its eighth edition, the USAR provides recommendations for how financial data is reported in a foodservice operation.

Figure 12.2 is an example of an income statement prepared utilizing the USAR recommended

Key Term

Income statement: Formally known as "The Statement of Income and Expense" is a report that summarizes a foodservice operation's profitability including details regarding revenue, expenses, and profit (or loss) incurred during a specific accounting period. Also commonly called the profit and loss (P&L) statement.

Key Term

Uniform system of accounts for restaurants (USAR): A recommended and standardized (uniform) set of accounting procedures used to categorize and report a foodservice operation's revenue and expenses.

Shondra's Restaurant

Income Statement

For the Year Ended December 31, 20xx

Line		
1	**SALES**	
2	Food	$ 1,891,011
3	Beverage	$ 415,099
4	**Total Sales**	**$2,306,110**
5	**COST OF SALES**	
6	Food	$ 712,587
7	Beverage	$ 94,550
8	**Total Cost of Sales**	$ 807,137
9	**LABOR**	
10	Management	$ 128,219
11	Staff	$ 512,880
12	Employee Benefits	$ 99,163
13	**Total Labor**	$ 740,262
14	**PRIME COST**	**$ 1,547,399**
15	**OTHER CONTROLLABLE EXPENSES**	
16	Direct Operating Expenses	$ 122,224
17	Music and Entertainment	$ 2,306
18	Marketing	$ 43,816
19	Utilities	$ 73,796
20	General and Administrative Expenses	$ 66,877
21	Repairs and Maintenance	$ 34,592
22	**Total Other Controllable Expenses**	$ 343,611
23	**CONTROLLABLE INCOME**	$ 415,100
24	**NON-CONTROLLABLE EXPENSES**	
25	Occupancy Costs	$ 120,000

Figure 12.2 Sample USAR Income Statement

26	Equipment Leases	$ 0
27	Depreciation and Amortization	$ 41,510
28	**Total Non-Controllable Expenses**	$ 161,510
29	**RESTAURANT OPERATING INCOME**	**$253,590**
30	Interest Expense	$ 86,750
31	**INCOME BEFORE INCOME TAXES**	**$166,840**

Figure 12.2   (Continued)

format. *Note*: The extreme left column (labeled "Line") has been added by the authors to make it easy to locate the various entries.

While **payroll** is the term commonly used to identify an operation's cash payments for labor, a USAR formatted income statement provides much greater labor cost-related details.

Note that "Labor" as shown in Figure 12.2 (Line 9) in a USAR-formatted income statement is separated into three distinct categories:

✓ Management
✓ Staff
✓ Employee Benefits

Key Term

Payroll: The term commonly used to indicate the amount spent for labor in a foodservice operation. Used, for example, in "last month our total payroll was $28,000."

Management (Line 10) includes the total dollar amount of salaries paid during the accounting period, and Staff (Line 11) refers to payments made to hourly (non-salaried) workers. The employee benefits category (Line 12) includes the cost of all benefits payments made for managers and hourly workers. Some benefit payments are mandatory (such as FICA [Social Security]), and others are voluntary (for example, the cost of providing health insurance).

Specific employee benefits paid by an operation vary but can include:

✓ (Social Security) taxes including taxes due on employees' tip income
✓ FUTA (federal unemployment taxes)
✓ SUTA (state unemployment taxes)
✓ Workers' compensation
✓ Group life insurance

Health insurance including:

- Medical
- Dental
- Vision
- Hearing
- Disability

✓ Pension/retirement plan payments
✓ Employee meals
✓ Employee training expenses
✓ Employee transportation costs
✓ Employee uniforms, housing, and other benefits
✓ Vacation/sick leave/personal days
✓ Tuition reimbursement programs
✓ Employee incentives and bonuses

Not every operation will incur all the benefit costs listed above, but some operations may have all (most) of these and perhaps even additional benefits.

In Figure 12.2, **Total Labor** (Line 13) is the sum of Lines 10, 11, and 12. The amount of labor cost is listed prominently on a USAR income statement because controlling and evaluating total labor cost is important in every foodservice operation.

Key Term

Total labor: The cost of the management, staff, and employee benefits expenses required to operate a foodservice when using a USAR formatted income statement.

"Total Labor" (Line 13) represents a key number for every foodservice operator seeking to better understand the complete cost of labor. However, it is not the only entry on a USAR income statement that must be understood when assessing labor costs. Important labor-related information exists in all three of the major segment sections of a USAR formatted income statement:

1) Sales
2) Expenses
3) Profits

Sales

In Figure 12.2, "Total Sales" of $2,306,110 are shown in Line 4. "Total Sales" is calculated when "Food" sales (Line 2) and "Beverage" sales (Line 3) are added together. It is easy to understand that an operation's sales level directly impacts its total labor costs.

Increased revenue, in almost all cases, will require increased labor-related expenditures, and this is expected. The most important "take away" when assessing an operation's total sales is that, while declining sales yield declining labor costs, this is certainly not desired! Almost always, a food service operator wants increasing levels of sales, even though these increases are also likely to increase the operator's cost of labor.

The amount of total sales is also an important number to know because one key assessment of labor productivity is the labor cost percentage (see Chapter 10). Calculating an accurate labor cost percentage requires that a foodservice operator knows the total sales generated during the accounting period for which the labor cost percentage is being calculated. This total sales number, the denominator (under the line) in the labor cost percentage formula, can be easily identified in a USAR-formatted income statement.

Expenses

In addition to listing its income, a USAR-formatted income statement details an operation's costs of doing business. These include the operation's "Total Cost of Sales (Line 8), Total Labor (Line 13), and its **prime cost** (the sum of Lines 8 and 13)" as shown on Line 14.

Prime cost is clearly listed on the income statement because it is an excellent indicator of an operator's ability to control product costs (cost of sales) *and* labor costs, the two largest expenses in most foodservice operations. The prime cost concept is also important because, when prime costs are excessively high, it can be very difficult to generate a sufficient level of profit in a foodservice operation, even when other costs are well-managed.

In addition to prime costs, the expenses section of a USAR-formatted income statement lists all of an operation's other costs of doing business. These include its **Total Other Controllable Expenses** (the sum of Lines 16–21 as shown in Line 22), and its **Total Non-Controllable Expenses** (the sum of Lines 25–27 as shown in Line 28).

Key Term

Prime cost: An operation's cost of sales plus its total labor costs.

Key Term

Total other controllable expenses: Expenses that a foodservice operator can influence with increases or decreases based on business decisions. Examples include marketing costs and utility costs.

Key Term

Total non-controllable expenses: Costs which, in the short run, cannot be avoided or altered by management decisions. Examples include equipment leases and occupancy payments such as leases and mortgages.

The importance of an operator's careful review of Total Other Controllable Expenses and all Total Non-controllable Expenses is easy to understand because the USAR format for an income statement does *not* include a separate line for many of an operation's actual "labor-related" costs. For example, the cost of "training materials" is clearly an employee-related expense, but the cost of those materials would be included in Line 20, "General and Administrative." Similarly, the cost of providing operation-specific uniforms (such as hats and shirts with an operation's name or logo) would be included in Line 16 "Direct Operating Expense" rather than in a unique expense category. Therefore, experienced food service operators recognize that "Total Labor" listed on a USAR income statement include the majority, but not all, of an operation's labor-related costs. Labor-related expenses may be recorded in several expense categories utilized by the USAR and, as a result, each must be carefully examined.

Profits

It is interesting that the word "profit" does not actually appear anywhere on a USAR income statement. Some foodservice operators consider "Restaurant Operating Income" (Line 29) to be their business's profit because it represents an operation's sales (revenue) minus all controllable and non-controllable expenses, and it reflects the basic profit formula:

$$Revenue - Expense = Profit$$

Other operators consider Income Before Income Taxes (Line 31) to be their profit, because it is calculated as Restaurant Operation Income minus Interest expense (Line 30).

Regardless of their preference for identifying profits, the most important points for foodservice operators to recognize concerning their labor-related costs and the income statement are that:

1) Increased sales will result in increased labor costs
2) A foodservice operation's "true" labor-related expense can include costs recorded in a variety of different expense categories
3) Excessive labor-related expense will cause profits to suffer

Foodservice operators can create their own income statements or rely on the services of a professional accounting firm. Regardless of the approach used, a basic understanding of the income statement is essential to a complete understanding of the management and control of labor costs and the optimization of foodservice profits.

Find Out More

Some foodservice operators do their accounting in-house without the assistance of outside accounting experts. However, properly accounting for a foodservice operation's revenue and expense is a very important task. This is especially the case when a foodservice operator seeks a bank loan or investment from outside investors, and, in these situations, an operation's accounting records most often must reflect the use of the Uniform System of Accounting for Restaurants (USAR).

Foodservice professionals working in the industry have a variety of tools available to assist with their accounting tasks. One of the newest and best is the book *Accounting and Financial Management in Foodservice Operations* written by Dr. David Hayes and Dr. Jack Ninemeier and published by John Wiley.

To review the contents of this book, prepared specifically for foodservice operators who want to utilize the USAR to create and evaluate their income statements (and more!), go to "www.wiley.com." When you arrive at the Wiley website, enter *Accounting and Financial Management in Foodservice Operations* in the search bar to examine the outline and content of this valuable accounting resource.

Managing and Controlling Total Labor Costs

The major components of a foodservice operation's total labor cost can be viewed in several ways. As you have read, "payroll" is often viewed as the key component of a foodservice operation's total labor costs. The reason: some employees are needed simply to open the doors for minimally anticipated business. For example, in a small ice-cream shop, payroll required to stay open may include only one worker. In another small operation, it may include only one supervisor, one server, and one cook. The cost to the operations of providing "payroll" in these two examples would be very small.

Assume, however, that a larger full-service restaurant anticipated much greater than minimum business volumes. The large number of expected guests requires that the operation will need more cooks, servers, cashiers, dish room personnel, and perhaps one or more supervisors to handle the additional workload. These additional positions create a work group that is far larger than the minimum staff required to open the doors of the operation but is needed to adequately serve the anticipated number of guests. In this example, payroll costs will be quite large. Regardless of their volume levels, all foodservice operators must manage and control their payroll costs if they are to optimize total labor expense.

Types of Labor Costs

The payroll costs in a foodservice operation may be viewed in several ways, including being either fixed or variable. **Fixed payroll** most often refers to what an operation pays in salaries. This amount is typically fixed because it remains unchanged from one pay period to the next pay period unless, for example, a salaried employee separates employment from the organization or is given a raise.

Variable payroll consists primarily of those dollars paid to hourly employees (staff) and is an amount that should vary with changes in sales volume. Generally, as sales volume increases, variable payroll expenses increase. In many cases, foodservice operators have little control over their fixed payrolls, but they have nearly 100% control over variable payroll expenses above their minimum staff levels.

Payroll costs may also be viewed as being either controllable or non-controllable. Payroll expenses that are wholly or partially controlled by management are considered **controllable labor expenses**, and labor costs beyond the direct control of management are **non-controllable labor expenses**. Examples of non-controllable labor expenses include employment taxes and some mandatory benefits. It is important that effective foodservice operators spend most of their time addressing their controllable labor expense rather than their non-controllable labor expense.

Key Term

Fixed payroll: The amount an operation pays for its salaried workers.

Key Term

Variable payroll: The amount an operation pays for workers compensated based on the number of hours worked.

Key Term

Controllable labor expenses: Those labor expenses that are under the direct influence of management.

Key Term

Non-controllable labor expenses: Those labor expenses that are not typically under the direct influence of management.

Controlling Total Labor Costs

Proper control and management of labor costs are important because a variety of laws regulate how and when employees must be paid. Many foodservice operators should pay attention to three key areas:

✓ Regular and overtime pay
✓ Accounting for tips
✓ Scheduling pay (where applicable)

Regular and Overtime Pay

In the United States, the Fair Labor Standards Act (FLSA; see Chapter 2) establishes minimum wage, overtime pay, recordkeeping, and youth employment standards affecting employees in all businesses employing two or more workers. Covered non-exempt workers are entitled to be paid no less than the Federal minimum per hour. Overtime pay, at a rate not less than one and one-half times the regular rate of pay, is required after 40 hours of work in a workweek. The FLSA sets worker pay requirements in several important areas:

Minimum wage. The FLSA establishes a minimum hourly rate that must be paid to all employees. If a foodservice operation is located in a state with a mandated minimum wage that is higher than the federal minimum wage, the state's minimum wage must be paid.

Overtime. Covered nonexempt employees must receive overtime pay for hours worked over 40 per workweek (any fixed and regularly recurring period of 168 hours – seven consecutive 24-hour periods) at a rate that is not less than one and one-half times the regular rate of pay.

There is no federal limit on the number of hours employees who are 16 years or older may work in any workweek. The FLSA does not require overtime pay for work on weekends, holidays, or regular days of rest unless overtime is worked on such days.

Hours worked. Hours worked ordinarily include the time during which an employee is required to be on the employer's premises, on duty, or at a prescribed workplace.

Recordkeeping. Employers must display an official poster outlining the requirements of the FLSA. Employers must also keep accurate employee time and pay records.

Find Out More

Every employer covered by the FLSA must keep certain records for each covered, non-exempt worker. There is no required form for the records, but the records must include accurate information about the employee and data about the hours worked and the wages earned. The following is a listing of the basic records that every employer must maintain:

✓ Employee's full name and social security number
✓ Address including zip code
✓ Birth date, if younger than 19
✓ Sex and occupation
✓ Time and day of week when employee's workweek begins

(Continued)

✓ Hours worked each day and total hours worked each workweek
✓ Basis on which employee's wages are paid
✓ Regular hourly pay rate
✓ Total daily or weekly straight-time earnings
✓ Total overtime earnings for the workweek
✓ All additions to or deductions from the employee's wages
✓ Total wages paid each pay period
✓ Date of payment, and the pay period covered by the payment

In addition to federal record keeping requirements, some states have their own payroll record keeping requirements. To learn more about payroll record keeping requirements in a specific state, enter "payroll record keeping requirements in (state name)" in your favorite search engine, and review the results.

Child labor. Child labor provisions set by the FLSA are designed to protect the educational opportunities of minors and prohibit their employment in jobs and under conditions detrimental to their health or well-being.

Every state has its own laws specifically dealing with child labor issues. When federal and state standards are different, the rules that provide the most protection to youth workers apply. U.S. employers must comply with both federal law and applicable state laws.

Accounting for Tips

Various methods can be used to account for employee tips, tip pooling, or tip sharing (see Chapter 10). However, regardless of the tip arrangement in place employers must carefully record employee tips and pay appropriate taxes on the tipped amounts. When tips are added to guests' bills that are paid with credit or with debit cards it is easy for an employer to record the amount of tips received. It is also important to recognize, however, that cash tips require applicable employer payments as well, and these amounts must be recorded as part of an operation's total labor costs.

Find Out More

Employees who receive cash tips of $20 or more in a calendar month while working are required to report the total amount of tips they receive to their employers. The employees must give their employers a written report by the tenth day of the following month, and this includes their cash tips. Cash tips include tips received directly from customers, tips from other employees under any tip-sharing arrangement, and charged tips (for example, credit and debit card charges) distributed to the employee.

Foodservice operators who receive tip reports from their employees use them to figure the amount of social security, Medicare, and income taxes to withhold for the pay period on both employee wages and reported tips. These expenses become part of an operator's total labor cost and the amount must be recorded for and reported on an operation's income statement.

IRS regulations in this area can be complex and are constantly changing. For example, in 2023, the IRS proposed a new Service Industry Tip Compliance Agreement (SITCA) program. SITCA is a voluntary tip reporting program for service industry employers (excluding gaming industry employers). The IRS intends to replace its existing tip reporting programs with SITCA which aims to:

1) Take advantage of new technologies (i.e., POS systems, time and attendance systems, and electronic payment settlement methods)
2) Increase tip reporting compliance
3) Decrease taxpayer and IRS administrative burden

To find out more about SITCA, enter "IRS SITCA program" in your favorite search engine and review the results.

Scheduling Pay (Where Applicable)

As addressed in Chapter 10, predictive scheduling regulations can impact a foodservice operator's total labor cost. While there has been little regulation requiring employers to provide advanced notice of schedule modifications in the past, increasingly, in some states and local jurisdictions, predictive scheduling legislation places limitations on how and when employers may modify a posted work schedule.

In some states employers are penalized if a scheduled worker is removed from the schedule without appropriate notice. In other cases, employees must be paid a minimum amount for reporting to work (and especially so when the number of hours they work is less than those originally posted on their schedule).

Some state and local scheduling laws require employers to pay a predictive scheduling penalty when the employer changes an employee's schedule without a loss in hours but also without the advance notice required by the state or local scheduling law. These payments are made above and beyond the employee's straight time or overtime earnings for the hours worked. Other states, in some circumstances, require employers to pay an employee who is scheduled for an **on-call** shift, but who is not ultimately called in to work.

Key Term

On-call (shift): An employee who is on-call and not working but who is required to be available in case they are needed to work.

Under regulations issued under the FLSA, employers must pay a non-exempt employee for on-call time if they "are required to remain on call at the employer's premises or so close thereto that they cannot use the time effectively for their own purposes."[1]

In those states where predictive scheduling regulations or other scheduling-related regulations are not in place, employers can modify their schedules as they wish. Of course, excessive modifications of schedules can have a negative impact on employee morale, and they should be avoided whenever possible.

What Would You Do? 12.1

"OK, explain the options to me one more time," said Micky Jospeh, the owner of the Hummus Hub, a small Lebanese shop that features a variety of types of kebabs and a wide selection of small dishes served as sides and appetizers.

Micky was talking with Guy Larue, the salesperson for the "Get-It-Now" third-party delivery company. Mickey did not offer delivery service, and most of his lunchtime business is from pick-up orders. Guests of the Hummus Hub either call in their orders or place their orders via text. After investigating the cost of providing delivery service to many of those who placed lunchtime orders, Mickey had contacted "Get-It-Now," a new third-party delivery company to learn about the services and charges they offer to small operations like the Hummus Hub.

"You have three choices," said Guy, "with our mini-plan we charge you a 15% fee for delivery orders, and it only shows your restaurant to guests searching for you by name. With our mini-Plus plan, you pay a 25% fee, and it shows your restaurant on our home screen and search results of the app. With our Premium-Plus program, you pay a 30% fee for delivery orders, and it shows your restaurant high-up on our home screen. The good news for you is that you will incur zero payroll and delivery costs with any plan you choose!"

Assume you were Micky. Do you agree that, in a partnership with "Get-It-Now," you would incur zero labor-related costs? Where would these "non-labor" delivery costs be recorded? Explain your answer.

Assessment of Total Labor Costs

To determine how much labor expense is needed to properly operate their businesses, foodservice operators must be able to determine how much work is to be done and how much work each employee can accomplish. If too few employees are scheduled to work, poor service and reduced sales can result because guests

---

1 https://webapps.dol.gov/elaws/whd/flsa/hoursworked/screenEr80.asp#:~:text=An%20 employee%20who%20is%20required,case%2Dby%2Dcase%20basis. Retrieved October 26, 2023.

may choose to go elsewhere in search of superior service levels. If too many employees are
scheduled, staff wages and employee benefits
costs will be higher than necessary, and the
result will be reduced profits.

To properly determine the number of workers
needed, operators must have a good understanding of the **productivity** level of each employee.

There are several ways to assess labor productivity, but productivity is generally measured by
calculating a **ratio.**

The formula used to calculate a productivity
ratio is:

Key Term

Productivity (worker): The
amount of work performed by
an employee within a fixed
time period.

Key Term

Ratio: An expression of the
relationship between two
numbers that is computed by
dividing one number by the
other number.

$$\frac{\text{Output}}{\text{Input}} = \text{Productivity Ratio}$$

To illustrate the use of this ratio, assume a foodservice operation employs 4 servers, and one day it served 80 guests. Using the above productivity ratio formula,
the output is guests served, and the input is servers employed:

$$\frac{80 \text{ guests}}{4 \text{ servers}} = 20 \text{ guests per server}$$

This formula states that, for each server employed, 20 guests were served. The
productivity ratio is 20 guests to 1 server (20 to 1) or, stated another way, 1 server
per 20 guests (1/20).

There are several ways of defining foodservice output and input, and there are
several types of productivity ratios. Productivity ratios can help an operator determine the answer to the key question, "How much should I spend on labor?" The
answer to the question becomes more complex, however, when one recognizes
that productivity levels in back-of-house areas (those unavailable for public
access) and front-of-house areas (those open to public access) should, in most
cases, be measured differently.

While each foodservice operation is unique, typical measures used to assess
back-of-house productivity include:

✓ Number of covers (guest orders) completed per labor hour
✓ Number of guest checks (table orders) processed per labor hour
✓ Average guest check completion time (in minutes)
✓ Number of menu items produced per hour worked
✓ Number of improperly cooked or incorrect items (mistakes) produced per
  hour worked

Examples of typical measures used to assess the <u>front-of-house</u> productivity of servers include:

✓ Number of guests (not tables) served per server hour worked
✓ Number of menu "specials" sold per server hour worked
✓ Number of errors (voided sales) produced per shift worked
✓ Average guest check size (per guest or table served)

Regardless of the productivity measures used, foodservice operators must develop their own methods for managing payroll costs because every foodservice unit is different. Consider, for example, the differences between managing payroll costs incurred by a food truck operator and those required for a large banquet kitchen located in a 1,000-room convention hotel.

Although methods used to manage payroll costs may vary, payroll costs can and must be managed. While several ratios are commonly used by operators assess their labor costs (worker productivity), three commonly utilized productivity ratios are:

1) Sales per Labor Hour
2) Labor Dollars per Guest Served
3) Labor Cost Percentage

Sales per Labor Hour

It has been said that the most perishable commodity any foodservice operator can buy is a labor hour. When labor is not productively used, it disappears forever. It cannot be "carried over" to the next day like an unsold head of lettuce or a slice of turkey breast. For this reason, some foodservice operators measure labor productivity in terms of the amount of sales generated for each labor hour used. This productivity measure is referred to as **sales per labor hour**.

Key Term

Sales per labor hour: The dollar value of sales generated for each labor hour used.

The formula used to calculate sales per labor hour is:

$$\frac{\text{Total Sales Generated}}{\text{Total Labor Hours Used}} = \text{Sales per Labor Hour}$$

As noted above, this productivity measure is the sum of all labor hours paid for by an operation within a specific sales period (Total Labor Hours Used) divided into the Total Sales Generated within the same period. To illustrate, consider the

Week	Total Sales Generated	Labor Hours Used	Sales per Labor Hour
1	$18,400	943.5	$19.50
2	21,500	1,006.3	21.37
3	19,100	907.3	21.05
4	24,800	865.3	28.66
Total	$83,800	3,722.4	$22.51

Figure 12.3 Four-Week Sales per Labor Hour

operator whose four-week labor usage and the resulting sales per labor hour information is presented in Figure 12.3.

In this example, sales per labor hour ranged from a low of $19.50 in Week 1 to a high of $28.66 in Week 4. *Note*: Sales per labor hour varies with changes in selling prices, but it will not vary based on changes in prices paid for labor. In other words, increases and decreases in the price paid per hour of labor will not affect this productivity measure.

As a result, a foodservice operation paying its employees an average of $15.00 per hour could, using this type of measure for labor productivity, have the same sales per labor hour as another foodservice operation that pays $20.00 for each hour of labor used. The result: the operator paying $15.00 per hour has paid less for an equally productive workforce when the sales generated per labor hour used are identical in the two units.

Many operators like utilizing the Sales per Labor Hour productivity measure because records on both the numerator (Total Sales Generated) and the denominator (Total Labor Hours Used) are readily available. However, depending on the record-keeping system employed, it may be more difficult to determine total labor hours used than total labor dollars spent. This is especially true when many managers and/or supervisors are paid by salary rather than by the hour. *Note*: It is an operator's choice whether both salaried workers and hourly paid staff are considered when computing an operation's overall sales per labor hour. As long as the operator is consistent with the choice that is made, sales per labor hour from different sales periods can be compared.

Labor Dollars per Guest Served

Some foodservice operators measure labor productivity in terms of the labor dollars spent for each guest served. This productivity measure is referred to as **labor dollars per guest served**.

Key Term

Labor dollars per guest served: The dollar amount of labor expense spent to serve each of an operation's guests.

Week	Cost of Labor	Guests Served	Labor Dollars per Guest Served
1	$ 7,100	920	$7.72
2	8,050	1,075	7.49
3	7,258	955	7.60
4	6,922	1,240	5.58
Total	$29,330	4,190	$7.00

Figure 12.4   Four-Week Labor Dollars per Guest Served

The formula used to calculate labor dollars per guest served is:

$$\frac{\text{Total Cost of Labor}}{\text{Total Number of Guests Served}} = \text{Labor Dollars per Guest Served}$$

To illustrate the use of this ratio, consider the operator whose four-week labor cost and the resulting sales per labor hour information are presented in Figure 12.4.

In this example, the labor dollars expended per guest served for the four-week period would be computed as:

$$\frac{\$29,330}{\$4,190} = \$7.00$$

Note that, in this example, for three weeks (weeks 1–3) the operator provided guests with more than $7.00 of guest-related labor costs per guest served. However, in the fourth week that amount fell to less than $6.00 per guest. This productivity measure, when averaged, can be useful for operators who find that their labor dollars expended per guest served are lower when their volume is high, and higher when their volume is low.

The utility of labor dollars per guest served is limited because it varies based on the price paid for labor. Unlike sales per labor hour, however, it is not affected by changes in menu prices.

Labor Cost Percentage

The most commonly used measure of employee productivity in the foodservice industry is the labor cost percentage (see Chapter 10). Recall from Chapter 10 that the formula used to calculate a labor cost percentage is:

$$\frac{\text{Total Cost of Labor}}{\text{Total Revenue Generated}} = \text{Labor Cost \%}$$

A labor cost percentage allows an operator to measure the relative cost of labor used to generate a known quantity of sales. It is important to realize, however, that different operators may choose slightly different methods of calculating this popular productivity measure.

Since a foodservice operation's total labor cost consists of management, staff, and employee benefit costs, some operators may calculate their labor cost percentage using only hourly staff wages, or staff wages and salary costs, but not benefit costs. This approach makes sense if an operator can directly control employee pay but not employee benefit costs. It is important to recognize, however, that when operators wish to directly compare their own labor cost percentage to that of other operations, both must have used the same labor cost percentage formula.

Controlling the labor cost percentage is extremely important in the foodservice industry because it is often used to assess management effectiveness. If an operation's labor cost percentage increases beyond what is expected, management will likely be held accountable by the operation's ownership.

Labor cost percentage is a popular measure of productivity, in part, because it is so easy to compute and analyze. To illustrate, consider Miriam, a foodservice manager in charge of a casual service restaurant in a year-round theme park. The restaurant is popular and has a $20 per guest check average. Miriam uses only payroll (staff wages and management salaries) when determining her overall labor cost percentage. The reason: she does not have easy access to the actual amount of taxes and benefits provided to her employees. Miriam's own supervisor considers these labor-related expenses to be non-controllable and, therefore, beyond Miriam's immediate influence.

Miriam has computed her labor cost percentage for each of the last four weeks using her modified labor cost percentage formula. Her supervisor has given Miriam a goal of 35% labor costs for the four-week period. Miriam feels that she has done well in meeting that goal. Figure 12.5 shows Miriam's four-week performance.

Week	Cost of Labor	Total Sales	Labor Cost %
1	$ 7,100	$18,400	38.6%
2	8,050	21,500	37.4
3	7,258	19,100	38.0
4	6,922	24,800	27.9
Total	**$29,330**	**$83,800**	**35.0**

Figure 12.5 Miriam's Four-Week Labor Cost Percentage Report

Using her labor cost percentage formula and the data in Figure 12.5, Miriam's Labor Cost % is calculated as:

$$\frac{\text{Cost of Labor}}{\text{Total Sales}} = \text{Labor Cost \%}$$

Or

$$\frac{\$29,330}{\$83,800} = 35\%$$

While Miriam did achieve a 35% labor cost for the four-week period, Mary, her supervisor, is concerned because she received several negative comments in week 4 regarding poor service levels in Miriam's unit. Some of these were even posted online, and Mary is concerned about the postings' potential impact on future visitors to the park's foodservice operations. When she analyzes the numbers in Figure 12.5, Mary sees that Miriam exceeded her goal of a 35% labor cost in weeks 1 through 3 and then reduced her labor cost to 27.9% in week 4.

Although the monthly overall average of 35% is within budget, Mary knows all is not well in this unit. To achieve her assigned goal, Miriam elected to reduce her payroll in week 4. However, the negative guest comments suggest that reduced guest service resulted from too few employees to provide the necessary guest attention. As Mary recognized, one disadvantage of using an overall labor cost percentage is that it can hide daily or weekly highs and lows.

In Miriam's operation, labor costs were too high the first three weeks, and too low in the last week, but she still achieved her overall target of 35%. Miriam's labor cost of 35% indicates that, for each dollar of sales generated, 35 cents was paid to the employees who assisted in generating those sales. In many cases, a targeted labor cost percentage is viewed as a measure of employee productivity and, to some degree, management's skill in controlling labor costs.

While it is popular, in addition to its tendency to mask productivity highs and lows, the labor cost percentage has other limitations as a measure of productivity. Note, for example, what happens to this measure of productivity if all of Miriam's employees are given a 5% raise in pay. If this were the case, her labor cost percentages for last month would be calculated as shown in Figure 12.6.

Note that labor now accounts for 36.8% of each sales dollar, but one should realize that Miriam's staff members did not become less productive simply because they got a 5% increase in pay. Rather, the labor cost percentage changed because of a difference in the price paid for labor. When the price paid for labor increases, the labor cost percentage increases and, when the price paid for labor decreases, the labor cost percentage decreases. Therefore, using the labor cost percentage alone to evaluate workforce productivity can sometimes be misleading.

Week	Original Cost of Labor	5% Pay Increase	Cost of Labor (with 5% Pay Increase)	Total Sales	Labor Cost %
1	$7,100	$355.00	$7,455.00	$18,400	40.5%
2	8,050	402.50	8,452.50	21,500	39.3
3	7,258	362.90	7,620.90	19,100	39.9
4	6,922	346.10	7,268.10	24,800	29.3
Total	29,330	1,466.50	30,796.50	83,800	36.8

Figure 12.6 Miriam's Four-Week Revised Labor Cost % Report (Includes 5% Pay Increase)

Another example of a limitation of the labor cost percentage as a measure of labor productivity can be seen when selling prices are increased. Return to the data in Figure 12.5 and assume that Miriam's unit raised all menu prices by 5% effective at the beginning of the month. Figure 12.7 shows how this increase in her selling prices would affect her labor cost percentage.

Note that increases in selling prices (when there is no decline in guest count or changes in guests' buying behavior) result in *decreases* in the labor cost percentage. Alternatively, lowering selling prices without increasing total revenue by an equal amount will result in *increases* in labor cost percentage.

Although labor cost percentage is easy to compute and widely used, it is difficult to use as a measure of productivity over time. The reason: it depends on labor dollars spent and sales dollars received for the computation. Even in relatively non-inflationary times, wages do increase, and menu prices are adjusted (usually upwards). Both activities directly affect the labor cost percentage, but not worker productivity. In addition, consider institutional foodservice settings that often have no daily dollar sales figures to report. These facilities can discover that it is not easy to measure labor productivity using labor cost percentages. The reason:

Week	Cost of Labor	Original Sales	5% Selling Price Increase	Sales (with 5% Selling Price Increase)	Labor Cost %
1	$7,100	$18,400	$920	$19,320	36.7%
2	8,050	21,500	1,075	22,575	35.7
3	7,258	19,100	955	20,055	36.2
4	6,922	24,800	1,240	26,040	26.6
Total	29,330	83,800	4,190	87,990	33.3

Figure 12.7 Miriam's Four-Week Revised Labor Cost % Report: (Includes 5% Increase in Selling Price)

operators generally calculate, and report guest counts or number of meals served rather than sales dollars earned.

Figure 12.8 summarizes key characteristics of the three measures of labor productivity presented.

Regardless of the productivity measure utilized, if an operator finds labor costs are too high relative to sales produced, problem areas must be identified, and corrective action(s) must be taken. If the overall productivity of employees cannot be improved, other action(s) become important.

The approaches operators can take to reduce labor-related costs are different for fixed payroll costs than for variable payroll costs. Figure 12.9 summarizes strategies operators can use to reduce labor-related expense percentages in each of these two categories. Note that operators can only decrease variable payroll expenses by increasing productivity, improving the scheduling process, eliminating employees, and/or reducing wages paid.

Another tactic that can increase employee productivity and reduce labor-related expense is employee empowerment (see Chapter 1). Employee empowerment results from a decision by management to fully involve employees in the decision-making process.

Measurement	Advantages	Disadvantages
Sales per Labor Hour $= \dfrac{\text{Total Sales Generated}}{\text{Total Labor Hours Used}}$	1) Fairly easy to compute 2) Does not vary with changes in the price of labor	1) Ignores price per hour paid for labor 2) Varies with changes in menu selling price
Labor Dollars per Guest Served $= \dfrac{\text{Total Cost of Labor}}{\text{Total Number of Guests Served}}$	1) Fairly easy to compute 2) Does not vary with changes in menu selling price 3) Can be used by non-revenue-generating units	1) Ignores average sales per guest and, therefore, total sales 2) Varies with changes in the price of labor
Labor Cost Percentage $= \dfrac{\text{Total Cost of Labor}}{\text{Total Revenue Generated}}$	1) Easy to compute 2) Most widely used	1) Hides highs and lows 2) Varies with changes in the price of labor 3) Varies with changes in menu selling price

Figure 12.8 Productivity Measures Summary

Labor Category	Actions
Fixed	1) Increase sales volume.
	2) Combine jobs to eliminate fixed positions.
	3) Reduce wages paid to fixed-payroll employees.
Variable	1) Improve productivity.
	2) Schedule appropriately to adjust to changes in sales volume.
	3) Combine jobs to eliminate variable positions.
	4) Reduce wages paid to variable employees.

Figure 12.9   Reducing Labor-Related Expense

Experienced foodservice operators remember that it was once customary for management to (a) make all decisions regarding every facet of the operational aspects of its organization and (b) present them to employees as facts to be accomplished by the staff members. Instead, an alternative approach occurs when employees are given the "power" to get involved.

Employees can be empowered to make critical decisions concerning themselves and, most importantly, the operation's guests. Many front-of-house employees work closely with guests, and numerous problems are more easily solved when employees are given the power to make it "right" for the guests. Successful operators often find that well-planned and consistently delivered training programs can be helpful. Also, empowered employees can yield a loyal and committed workforce that is more productive, is supportive of management, and will "go the extra mile" for guests. Doing so helps reduce labor-related costs, builds repeat sales, and increases profits.

Technology at Work

Managing productivity levels in a foodservice operation is easier than ever today because of the large number of labor management software packages on the market. The best of such packages allow operators to:

1) Streamline scheduling to consider employee availability, preferences, and skill sets with a business's needs and peak hours.
2) Ensure legal compliance with labor laws and regulations including minimum wage requirements, overtime rules, and employee break policies.

(Continued)

3) Integrate labor scheduling with other management systems such as an operation's point-of-sale (POS) system, inventory management software, and customer relationship management (CRM) tools.
4) Obtain valuable insights and predictions about labor needs based on historical data and trends that enable operators to make more informed decisions about optimal staffing levels.

To find out more about the features of popular labor cost management tools for foodservice operators, enter "labor cost management software for restaurants" in your favorite search engine and view the results.

What Would You Do? 12.2

"I'm just saying that we are already reporting our daily labor cost percentage to the owners, and now they also want to know this every day. It just seems like more paperwork to me," said Larry Sullivan, the assistant manager of the Philadelphia Sliders restaurant.

Larry was talking to Talisha, the restaurant's manager. Talisha had just informed Larry that the restaurant's owners wanted her to submit a daily "Average Drive-thru Ticket Time Report" along with the day's revenue and labor cost percentage reports.

Along with on-site dining, the Philadelphia Sliders drive-through was seeing an increasing amount of business. However, there have been some complaints about slow service. In response, the restaurant's owners requested that Talisha and applicable staff calculate and report an average drive-thru ticket time. *Note*: Ticket time begins when a guest's order is placed, and it ends when the guest leaves the drive-thru window with their ordered items.

The restaurant's POS system was programmed to create a unique guest check for each order, and the system recorded the time at which a guest's order was placed and when the order was completed. One reporting feature of the POS system was that it automatically calculated the average drive-thru ticket time for all drive-thru orders, and this was the information the operation's owners now wanted to review each day.

Assume you were Talisha. Why do you think this operation's owners are now requesting this new employee productivity report? How important do you believe the information contained in this report is? Will this data help your labor cost control activities, and your efforts to optimize guest service in the drive-thru?

Key Terms

Split-shift	Total other controllable	Non-controllable labor
Income statement	expenses	expenses
Uniform system of	Total non-controllable	On-call (shift)
accounts for	expenses	Productivity (worker)
restaurants (USAR)	Fixed payroll	Ratio
Payroll	Variable payroll	Sales per labor hour
Total labor	Controllable labor	Labor dollars per
Prime cost	expenses	guest served

Operator's 10-Point Tactics for Success Checklist

Evaluate your need for, and the current status of, each of the following operational tactics. For those tactics you think are important, but not yet in place, develop an action plan for its implementation including who will be responsible for the tactic's completion and the target date by which it should be completed.

				If Not Done	
Tactic	**Don't Agree (Not Done)**	**Agree (Done)**	**Agree (Not Done)**	**Who Is Responsible?**	**Target Completion Date**
1) Operator understands the relationship between operating profits and the control of labor-related costs.	____	____	____		
2) Operator can identify important non-cash payment factors that will affect total labor costs.	____	____	____		
3) Operator can explain the three components that yield "Total Labor" when producing an income statement utilizing the USAR format.	____	____	____		
4) Operator understands the difference between fixed payroll and variable payroll.	____	____	____		

(Continued)

Tactic	Don't Agree (Not Done)	Agree (Done)	Agree (Not Done)	If Not Done	
				Who Is Responsible?	Target Completion Date
5) Operator recognizes the difference between controllable and non-controllable labor expenses.	——	——	——		
6) Operator can describe special issues to consider when accounting for regular and overtime pay, tips, and scheduling pay (where applicable).	——	——	——		
7) Operator knows how to calculate and assess the "Sales per Labor Hour" measure of productivity.	——	——	——		
8) Operator knows how to calculate and assess the "Labor Dollars per Guest Served" measure of productivity.	——	——	——		
9) Operator knows how to calculate and assess the "Labor Cost Percentage" measure of productivity.	——	——	——		
10) Operator can state the various actions that can be taken to help reduce fixed and variable labor-related expenses.	——	——	——		

Glossary

Absolute standard (performance appraisal method): Measuring an employee's performance against a previously established standard.

Accountability: The term used to indicate that employees take responsibility for both their performance and the results of that performance.

Adaptation (to an organization): The process by which new employees learn the values of and "what it is like" to work for a hospitality organization during their initial job experiences.

Americans with Disabilities Act (ADA): A federal law prohibiting discrimination against persons with disabilities.

Applicants: Job searchers who have officially "applied" for a job opening.

Application: A form, questionnaire, or similar document that applicants for employment are required by an employer to complete. An application may exist in a hard copy, electronic copy, or Internet medium.

Appraisal (employee): An objective and comprehensive evaluation of employee effectiveness.

Asset (business): Property that is used in the operation of a business including money, real estate, buildings, inventories, and equipment.

At-will (employment): An employment relationship in which an employer can, at any time, terminate the relationship without liability as long as it is not discriminatory. Likewise, an employee is free to leave a job at any time for any or no reason with no adverse legal consequences.

Back-of-house staff: The employees of a foodservice operation whose duties do not routinely put them in direct contact with guests.

Benchmark: A standard or point of reference used to measure something else.

Bona fide occupational qualification (BFOQ): A specific job requirement for a particular position that is reasonably necessary to the normal operation of a business, and therefore allows discrimination against a protected class (for example, choosing a female model when photographing an advertisement for lipstick).

Career development program: A planning strategy in which one identifies career goals and then plans education and training activities designed to attain them.

Child labor: Work that deprives children of their childhood, their potential, and their dignity, and that is harmful to physical and mental development.

Coaching: A training and supervisory activity that involves informal on-the-job conversations and/or demonstrations that encourage proper (and discourage improper) work behavior.

Code of ethics: A guiding set of principles intended to instruct professionals to act in a manner that is honest and beneficial to all an organization's stakeholders.

Compensation: Any form of cash and non-cash payment given to an individual for services rendered as an employee by their employer.

Compensation management (program): The process of administrating a foodservice operation's extrinsic and intrinsic reward systems.

Compensation package: The sum total of the money and other valuable items given to an employee in exchange for work performed.

Controllable labor expenses: Those labor expenses that are under the direct influence of management.

Core values: Deeply ingrained principles that guide an organization's actions. They are fundamental beliefs about a foodservice operation that dictate behavior and help its members understand what is considered right or wrong.

Cost-effective (training): A term that indicates that an item or activity such as training is worth more than it costs to provide it.

Crises (work): Situations that have the potential to negatively affect the health, safety, or security of employees.

Cross-functional teams: Groups of employees from different operating teams within a foodservice operation who work together to resolve problems.

Cross-training: A training tactic that allows employees to learn how to perform tasks in another position.

Delegate: To assign authority to subordinates to do work and make decisions normally made by someone higher at a higher organizational level.

Digital menu: An integrated system that uses hardware and software to display an operation's menu on an electronic screen; also commonly referred to as a digital display menu or digital menu board.

Discipline: Any effort designed to influence an employee's negative behavior.

Disciplined (work team): The situation in which employees conduct themselves according to accepted rules and standards of conduct.

Dismissal: An employer-initiated separation of employment.

Distance education: An individual training method in which a staff member enrolls in a for-credit or not-for-credit program offered by an educational facility or a professional association.

Diversity: The term used to describe the variety of differences between people in an organization that includes race, gender, age, religion, sexual orientation, and more.

Documented oral warning: The first step in a progressive discipline process: a written record of an oral reprimand given to an employee is provided.

E-learning: Learning conducted via electronic media and typically on the internet. Also commonly referred to as "online learning."

Employee assistance programs (EAPs): The term that describes a variety of employer-initiated efforts to assist employees in the areas of family concerns, legal issues, financial matters, and health maintenance.

Employee benefits: Any form of rewards or compensation provided to employees in addition to their base salaries and wages. Benefits offered to employees may be legally mandatory or provided voluntary.

Employee handbook: A permanent reference guide for employers and employees that contains information about a company, its goals, and its current employment policies and procedures. Also commonly referred to as an "Employee manual."

Employee referral: A promote-from-within recruitment approach used to identify qualified job applicants.

Employee turnover rate: The proportion of total employees in an operation replaced during a specific time period.

Employer of choice: A company whom workers choose to work for when presented with other employment choices. This choice is a conscious decision made when joining a company and when deciding to stay with the employer.

Employers' experience rating: An assessment of unemployment insurance taxes paid by employers who have paid covered wages for a sufficient period to rate their experience with unemployment insurance. In most cases, the less unemployment an employer's workers have experienced, the lower the employer's unemployment insurance tax rate will be.

Employment agreement: The terms of the employment relationship between an employer and employee that specifies the rights and obligations of each party.

Employment law: The body of laws, administrative rulings, and precedents that address the legal rights of workers and their employers.

Empowerment (employee): An operating philosophy that emphasizes the importance of allowing employees to make independent operations-related decisions.

Equal Employment Opportunity Commission (EEOC): The entity within the federal government assigned to enforce the provisions of Title VII of the Civil Rights Act of 1964.

Ethics: Standards that help decide between right and wrong and influence behavior toward others.

Exempt employee: An employee who is not subject to the minimum wage or overtime provisions of the Fair Labor Standards Act (FLSA).

Exit interview: A formal meeting between a representative of a foodservice operation and a departing employee.

Explicit threat: A threatening act that is fully and clearly expressed or demonstrated, leaving nothing implied. Example: The statement "If I see you in my work area again, I'll personally throw you out of it!" said in a menacing voice by one employee to another.

External customers: Guests of a foodservice operation.

External search: An approach to seeking job applicants that focuses on candidates not currently employed by the organization.

Extrinsic rewards: Financial and non-financial compensation granted to a worker by others (usually the employer).

Fixed payroll: The amount an operation pays for its salaried workers.

Flat rate (income tax): A taxing system in which each taxpayer pays the same percentage of their income in taxes.

Form I-9: Formally the "Employment Eligibility Verification," this document is a United States Citizen and Immigration Services form required for use by all employers. It is used to verify the identity and legal authorization to work of all paid employees.

401-K: A retirement program where an employee can make contributions from their paycheck either before or after-tax, depending on the options offered in the plan. If they desire to do so, employers can also make contributions to a worker's 401-k account.

Front-of-house staff: The employees of a foodservice operation whose duties routinely put them in direct contact with guests.

Full-time equivalent (employee): A unit of measurement used to calculate the number of full-time hours worked by all employees in a business. For example, if a business considers 40 hours to be a full-time workweek, then an employee working 40 hours per week would have an FTE of 1.0. In contrast, a part-time employee working only 20 hours per week would have an FTE of 0.5-which shows that the hours they worked are equivalent to one half of a full-time employee.

Garnish(ment): A court-ordered method of debt collection in which a portion of a worker's income is paid directly to one or more of that worker's creditors.

Good cause: A reason for taking an action or failing to take an action that is reasonable and justified when viewed in the context of all surrounding circumstances.

Good faith: A concept that requires parties to an agreement to exercise their powers reasonably and not arbitrarily or for some irrelevant purpose. Certain conduct may lack good faith if one party acts dishonestly or fails to have concern for the legitimate interests of the other party.

Graduated rate (income tax): A taxing system in which higher tax rates are applied to higher income workers.

Group training: A training method that involves presenting the same information to more than one trainee at the same time.

Halo effect: The tendency to let the positive assessment of one individual trait influence the evaluation of other traits that are unrelated.

Hostile work environment: A workplace infused with intimidation, ridicule, and insult that is severe or pervasive enough to create a seriously uncomfortable or abusive working environment with conduct severe enough to create a work environment that a reasonable person would find intimidating (hostile).

Implicit threat: A threatening act that is implied rather than expressly stated. Example: The statement "I'd watch my back if I were you!" said in a menacing voice by one employee to another.

Income statement: Formally known as "The Statement of Income and Expense" is a report that summarizes a foodservice operation's profitability including details regarding revenue, expenses, and profit (or loss) incurred during a specific accounting period. Also commonly called the profit and loss (P&L) statement.

Internal customers: Employees of a foodservice operation.

Internal search: A promote-from-within recruitment approach used to identify qualified job applicants.

Interstate commerce: Commercial trading or the transportation of people or property between or among states.

Intrinsic rewards: Self-initiated compensation (e.g., pride in one's work, a sense of professional accomplishment, and/or enjoying being part of a work team).

Involuntary (separation): An employer-initiated termination of employment.

Job description: A statement that outlines the specifics of a particular job or position within a foodservice operation. It provides details about the responsibilities and conditions of the job.

Job enlargement: An individual training method that involves adding tasks to a position that are traditionally performed at a higher organizational level.

Job enrichment: An individual training method that occurs when additional tasks that are part of a position at the same organizational level are added to another position at the same organizational level.

Job offer letter: A proposal by an employer to a potential employee that specifies employment terms. A legally valid acceptance of the offer creates a binding employment agreement.

Job rotation: The temporary assignment of employees to different positions or tasks to provide work variety or experience and to create "back-up" expertise within the organization.

Job specification: A listing of the personal characteristics and skills needed to perform the tasks contained in a job description.

Jurisdiction: The geographic area within which a court or government agency may legally exercise its power.

Just cause: A legally sufficient reason. Also sometimes referred to as good cause, lawful cause, or sufficient cause.

Labor cost percentage: A ratio of overall labor costs incurred relative to total revenue generated.

Labor dollars per guest served: The dollar amount of labor expense spent to serve each of an operation's guests.

Labor union: An organization that acts on behalf of its members to negotiate with management about the wages, hours, and other terms and conditions of the membership's employment.

Living wage: The minimum hourly wage necessary for a person to achieve a subjectively defined standard of living. In the context of developed countries such as the United States, this standard generally expresses that a person working 40 hours per week with no additional income should be able to afford a specified quality or quantity of housing, food, utilities, transportation, and healthcare.

Management: The process of planning, organizing, directing, controlling, and evaluating the financial, physical, and human resources of an organization to reach its goals.

Management by objectives (MBO): A plan developed by an employee and their supervisor that defines specific goals, tactics to achieve them, and corrective actions, if needed.

Mandatory benefit: An employee benefit that must, by law, be paid by an employer.

Master schedule: A listing of all employee shifts including specific start and end times for each shift during a calendar week.

Minimum wage: The lowest wage per hour that a worker may be paid as mandated by law.

Mission statement: A concise explanation of the organization's reason for existence that describes the organization's purpose and its overall intention.

Morale: The feelings including outlook, satisfaction, attitude, and confidence that team members have at work.

Mystery shopper: A person posing as a foodservice guest who observes and experiences an operation's products and services during a visit and who then reports findings to the operation's owner. Also referred to as a "secret shopper."

Negative discipline: Any action designed to correct undesirable employee behavior.

Negligent hiring: Failure of an employer to exercise reasonable care in the selection of employees.

Negligent retention: Retaining an employee after the employer becomes aware of an employee's unsuitability for a job by failing to act on that knowledge.

Non-controllable labor expenses: Those labor expenses that are not typically under the direct influence of management.

Objective test: An assessment tool such as a multiple choice or true/false test whose questions have only one correct answer and yield a reduced need for trainers to interpret trainees' responses.

Occupational Safety and Health Act: The law that gave the Federal Government the authority to establish and enforce safety and health standards for most employees in the United States.

Occupational Safety and Health Administration (OSHA): The agency of the Federal Government that assures safe and healthful working conditions by establishing and enforcing standards, and by providing necessary training, outreach, education, and assistance.

Off-the-shelf (training programs): Generic training materials typically addressing general topics of interest to many trainers that can be used if operation-specific resources have not been developed.

Onboarding: The process in which new team members are integrated into an organization. It includes activities that allow new employees to complete an initial new-hire orientation process, as well as learn about the organization and its structure, mission, and values.

On-call (shift): An employee who is on-call and not working but who is required to be available in case they are needed to work.

On-job training (method): An individualized training method in which a knowledgeable and skilled trainer teaches a less experienced staff member how to properly perform job tasks.

Open-door communication (policy): A policy that indicates to employees that a supervisor or manager is open to discuss an employee's questions, complaints, suggestions, and concerns.

Ordinance: A piece of legislation enacted by a municipal authority.

Organizational chart: A diagram that visually conveys a company's internal structure and its reporting relationships.

Orientation (employee): The process of providing basic information about a foodservice operation that every new employee within the facility should know.

Orientation (program): A carefully developed agenda of activities and content designed to train and enlighten new employees about their roles and company policies.

Orientation kit: A package of written materials given to new employees to supplement the oral information provided during an orientation session.

Paid time off (PTO) program: An employer-provided benefit in which an employee is allotted an amount of paid time that can be used for vacation, sickness, or personal time at their discretion.

Pay range: The span between the minimum and maximum hourly rate an organization will pay for a specific job or group of jobs.

Payroll: The term commonly used to indicate the amount spent for labor in a foodservice operation. Used, for example, in "last month our total payroll was $28,000."

Performance-based (training): A systematic way of organizing training to help trainees learn the tasks that are essential for effective on-job performance.

Performance management: A systematic process by which foodservice operators can help employees improve their ability to achieve goals.

Performance standards: Measurable quality and/or quantity indicators that identify when an employee is performing a task correctly.

Perpetual inventory: An inventory control system in which additions to and deletions from the total inventory are recorded as they occur.

Personal file: An individual worker's file that includes the worker's application for employment, and records which are used or have been used to determine the worker's qualifications for promotion, compensation, and/or termination. Information about disciplinary actions, if any, are also included in the personal file.

Personal protective equipment (PPE): Equipment worn to minimize exposure to hazards that can cause serious workplace injuries and illnesses.

Physical inventory: A tool in an inventory control system in which an actual (physical) count and valuation of all product inventory on hand is taken at the close of an accounting period.

Pitchfork effect: The tendency to allow the negative assessment of one individual trait to influence the evaluation of other, nonrelated traits.

Point-of-sale (POS) system: An electronic system that records foodservice customer purchases and payments, as well as other operational data.

Policy: A detailed course of action developed to guide future decision making to achieve stated objectives.

Position analysis: A process that examines each task listed for a position and explains how it should be done, with a focus on required knowledge and skills.

Positive discipline: Any action designed to encourage desired employee behavior.

Predictive scheduling legislation: Laws addressing the minimum amount of advance notice of work schedules that employers must provide to their employees.

Pretest/post-test evaluation: A before and after assessment used to measure whether the expected changes took place in a trainee's knowledge, skill level, and/or attitude after the completion of training.

Prime cost: An operation's cost of sales plus its total labor costs.

Procedure: A technique or method used to implement a policy.

Productivity (worker): The amount of work performed by an employee within a fixed time period.

Progressive discipline program: Program designed to modify employee behavior through a series of increasingly severe punishments for unacceptable behavior.

Protected class: A group of people who have special legal protection against discrimination in the workplace based on specifically identified traits.

QR code: A QR (quick response) code is a machine-readable bar code that, when read by the proper smart device, allows foodservice guests to view an online menu. Guests can also be re-directed to an online ordering website or an app that allows them to order and/or pay for their meal without having to interact directly with a staff member.

Quid pro quo (sexual harassment): Quid pro quo literally means "something for something." This harassment occurs when a supervisor behaves in a way or demands actions from an employee that force the employee to decide between giving in to sexual demands or losing their job, losing job benefits or promotion, or otherwise suffering negative consequences.

Ratio: An expression of the relationship between two numbers that is computed by dividing one number by the other number.

Reasonable accommodation: A change to the application or hiring process, to the job, to the way the job is one, or the work environment that allows a person with a disability who is qualified for the job to perform the essential functions of that job and enjoy equal employment opportunities.

Reasonable person standard: The typical or average person (and their behavior and beliefs) placed in a specific environmental setting.

Recruiting: The process of actively seeking out, finding, and hiring candidates for a specific position or job.

Reference (employment): The positive or negative comments about an employee's job performance provided to a prospective employer.

Relative standard (performance appraisal method): Measuring one employee's performance against another employee's performance.

Reliability (test): The ability of a test to yield consistent results.

Retaliation: Any act that results in an employer punishing an employee for advocating personal rights to be free from employment discrimination. These include a discriminatory workplace culture, violations of laws intended to protect health and safety, or acting as a whistleblower.

Role playing (exercise): An activity that allows trainees to act out a specific situation or assume the role of another person in a real-to-life situation or scenario.

Safe leave: Absence from work when an employee or employee's family member has been the victim of domestic abuse or violence.

Safety (employee): A condition that minimizes the risk of harm to employees.

Safety data sheets (SDSs): Documents that describe how to safely use, store, and dispose of potentially hazardous chemicals.

Salaried employee: An employee who regularly receives a predetermined amount of compensation each pay period on a weekly or less-frequent basis. The predetermined amount paid is not reduced because of variations in the quality or quantity (amount of time) that a salaried employee works.

Sales per labor hour: The dollar value of sales generated for each labor hour used.

Search engine results page (SERP): The web page of a search engine such as Google (used by Apple), Bing, or Yahoo shows a user when a reader types in a search query.

Security (worker): Employees' feelings about fear and anxiety.

Self-directed work teams: Groups of employees who combine their talents to work without the influence of traditional manager-based supervision.

Sexual harassment: Unwelcome sexual advances, requests for sexual favors, and other verbal or physical conduct of a sexual nature.

Split-shift: A working schedule comprising two or more separate periods of duty in the same day.

Stakeholders: Groups, individuals, and organizations that have a vested interest in the decision-making and activities of a business or organization.

Standardized recipe: The instructions needed to consistently prepare a specified quantity of food or drink at the expected quality level.

Standard operating procedure (SOP): The term used to describe the way something should be done under normal business operating conditions.

Step cost: A cost that increases in a non-linear fashion as activity or sales volume increases.

Suspension: The third step in a progressive discipline process; a (typically unpaid) period off from work resulting from on-going inappropriate behavior.

Targeted outcomes (performance appraisal method): Measuring the extent to which specified performance goals were achieved.

Task list: A detailed list of all tasks in a position.

Time and temperature control for safety food (TCS food): Foods that must be kept at a particular temperature to minimize the growth of food poisoning bacteria or to stop the formation of harmful toxins.

Tip credit: A legally permissible system that allows employers to include tips as part of a worker's wage calculation. Employers can credit a portion of an employee's received tips towards their minimum wage payment obligations so that employers do not have to pay the full minimum wage.

Tip distribution program: A system of tip payment that allows an operation to distribute a customer's tip from an employee who actually received it to others who also provided service to the customer.

Tip pooling: A tip system that takes the tips given to individual employees in a group and shares them equally with all other members of the group. Used for example, when bartender tips given to an individual bartender are shared equally with all bartenders working the same shift.

Tip sharing: A tip system that takes the tips given to one group of employees and provides a portion of them to another group of employees. Used, for example, when server tips are shared with those who bus the server's tables.

Title VII: A federal employment law that prohibits employment discrimination based on race, color, religion, sex, or national origin.

Total labor: The cost of the management, staff, and employee benefits expenses required to operate a foodservice when using a USAR formatted income statement.

Total non-controllable expenses: Costs which, in the short run, cannot be avoided or altered by management decisions. Examples include equipment leases and occupancy payments such as leases and mortgages.

Total other controllable expenses: Expenses that a foodservice operator can influence with increases or decreases based on business decisions. Examples include marketing costs and utility costs.

Training: The process of developing a staff member's knowledge, skills, and attitudes necessary to perform tasks required for a position.

Training handbook: A hard copy or electronic manual containing the training plan and associated training lessons for a foodservice operation's complete training program.

Training lesson: Information about a single session of a training plan. It contains one or more training objectives and indicates the content and method(s) to enable trainees to master the content.

Training plan: A description of the overview and sequence of a complete training program.

Train-the-trainer (program): A training framework that turns employees into subject matter experts who can effectively teach others.

Undue hardship: Action requiring significant difficulty or expense when considered in light of a number of factors. These factors include the nature and cost of the accommodation in relation to the size, resources, nature, and structure of the employer's operation.

Unemployment compensation benefits: Benefits paid to employees who involuntarily lose their employment without just cause.

Unemployment rate: The number of unemployed people in a community or other designated area who are seeking work; expressed as a percentage of the area's entire labor force.

Uniform system of accounts for restaurants (USAR): A recommended and standardized (uniform) set of accounting procedures used to categorize and report a foodservice operation's revenue and expenses.

User-generated content (UGC) site: A website in which content that includes images, videos, text, and audio have been posted on-line by the site's visitors. Examples of currently popular UGCs include Instagram, X (formerly Twitter), Threads, Facebook, Pinterest, and YouTube.

Validity (test): The ability of a test to evaluate only what it is supposed to evaluate.

Value: The amount paid for a product or service compared to the buyer's view of what they received in return.

Variable cost: An expense that generally increases as sales volume increases and decreases as sales volume decreases.

Variable payroll: The amount an operation pays for workers compensated based on the number of hours worked.

Voluntary (separation): An employee-initiated termination of employment.

Voluntary benefit: An employee benefit paid at the discretion of an employer in efforts to attract and keep the best possible employees.

Warm body syndrome: A selection error that involves hiring the first person who applies for a vacant position.

Wellness program (employee): An employer-sponsored initiative designed to promote the good health of employees.

Workers' compensation fund: A state-operated program that provides medical expenses, lost wages, and rehabilitation costs to employees who are injured or become ill "in the course and scope" of their job. It also pays death benefits to families of employees who are killed on the job. Payments are made into the fund by state-mandated employer premiums.

Workplace violence: Any act in which a person is abused, threatened, intimidated, or assaulted in their place of employment.

Written warning: The second step in a progressive discipline process that alerts an employee that further inappropriate behavior will lead to suspension.

Zero-tolerance (policy): A policy that permits no amount of leniency regarding harassing behavior.

Index